T0178237

SpringerBriefs in Mathematics

SpringerBriefs in Mathematics showcases expositions in all areas of mathematics and applied mathematics. Manuscripts presenting new results or a single new result in a classical field, new field, or an emerging topic, applications, or bridges between new results and already published works, are encouraged. The series is intended for mathematicians and applied mathematicians. All works are peer-reviewed to meet the highest standards of scientific literature.

More information about this series at http://www.springer.com/series/10030

Étienne Pardoux

Stochastic Partial Differential Equations

An Introduction

 Springer

Étienne Pardoux
Institut de Mathématiques de Marseille
Aix Marseille Université, CNRS
Marseille, France

ISSN 2191-8198 ISSN 2191-8201 (electronic)
SpringerBriefs in Mathematics
ISBN 978-3-030-89002-5 ISBN 978-3-030-89003-2 (eBook)
https://doi.org/10.1007/978-3-030-89003-2

Mathematics Subject Classification: 60H15, 35R60, 60G35, 60H07, 60J68

This Springer imprint is published by the registered company Springer Nature Switzerland AG
The registered company address is: Gewerbestrasse 11, 6330 Cham, Switzerland

Foreword

There is by now a growing interest in Stochastic Partial Differential Equations (abbreviated from now on as SPDEs). One can find two reasons for this.

First, more and more complex mathematical models are used in the applied sciences, in order to describe reality. The huge progress in computer power and our ability to simulate high-dimensional dynamical systems has made it possible to assess highly complex models, which take into account both randomness and the fact that most systems are distributed over space. This leads naturally to PDEs with random coefficients, and SPDEs.

Second, in recent decades we have seen the emergence of new sophisticated mathematical techniques, which allow us to tackle new problems and classes of equations. These include the theory of Rough Paths, first introduced by T. Lyons, see the course by Hairer and Friz [11], the theory of regularity structures invented by M. Hairer [10], and the method of paracontrolled distributions, due to Gubinelli, Imkeller and Perkowski [8].

The aim of these notes is to present a concise introduction to the "classical theory" of SPDEs, as it was developed during the last 25 years of the last century. We believe that a good understanding of this theory is useful in order to study and understand the new approaches.

Marseille, September 2021   *Étienne Pardoux*

Contents

Chapter 1
Introduction and Motivation

1.1 Introduction

In these lectures we shall study stochastic parabolic PDEs, most of which will be nonlinear. The general type of equations which we have in mind are of the form

$$\frac{\partial u}{\partial t}(t,x) = F(t,x,u(t,x),Du(t,x),D^2u(t,x)) + G(t,x,u(t,x),Du(t,x))\mathring{W}(t,x),$$

or in the *semilinear* case

$$\frac{\partial u}{\partial t}(t,x) = \Delta u + f(t,x,u(t,x)) + g(t,x,u(t,x))\mathring{W}(t,x).$$

We shall make precise what we mean by $\mathring{W}(t,x)$. We distinguish two cases

1. $\mathring{W}$ is white noise in time and colored noise in space. A particular case is that where the noise is of the form $\sum_{k=1}^{N} e_k(x)\mathring{W}_k(t)$.
2. $\mathring{W}$ is white noise both in time and in space.

In both cases, we can define $\mathring{W}$ in the distributional sense, as a centered generalized Gaussian process, indexed by test functions $h : \mathbb{R}_+ \times \mathbb{R}^d \mapsto \mathbb{R}$:

$$\mathring{W} = \{\mathring{W}(h); \ h \in C^\infty(\mathbb{R}_+ \times \mathbb{R}^d)\},$$

whose covariance is given by

$$\mathbb{E}\left(\mathring{W}(h)\mathring{W}(k)\right) = \int_0^\infty dt \int_{\mathbb{R}^d} dx \int_{\mathbb{R}^d} dy \, h(t,x)k(t,y)\varphi(x-y) \ \text{ in case 1}$$

$$= \int_0^\infty dt \int_{\mathbb{R}^d} dx \, h(t,x)k(t,x) \ \text{ in case 2.}$$

Here φ is a "reasonable" kernel, which might blow up to infinity at 0. Note that the first formula converges to the second one, if we let φ converge to the Dirac mass at 0. On the other hand, the solution of a PDE of the form

© The Author(s), under exclusive license to Springer Nature Switzerland AG 2021
É. Pardoux, *Stochastic Partial Differential Equations*,
SpringerBriefs in Mathematics, https://doi.org/10.1007/978-3-030-89003-2_1

$$\frac{\partial u}{\partial t}(t,x) = \Delta u(t,x) + f(t,x,u(t,x))$$

can be considered either as

1. a function of t with values in an infinite-dimensional space of functions of x (typically a Sobolev space); or else as
2. a real-valued function of (t,x).

Likewise, in the case of an SPDE of one of the above types, we can consider the solution either as

1. a stochastic process indexed by t, and taking values in an infinite-dimensional function space, i.e. the solution of an infinite-dimensional SDE; or else as
2. a one-dimensional random field, i.e. the solution of a multiparameter SDE.

The first point of view will be presented in Chapter 2. It applies mainly to equations driven by Gaussian noises which are colored in space. The second one will be presented in Chapter 3 for the study of space-time white noise driven SPDEs.

There are several serious difficulties in the study of SPDEs, which are due to the lack of regularity with respect to the time variable (resp. with respect to both the time and the space variable), and the interaction between the regularity in time and the regularity in space. As a result, as we will see, the theory of nonlinear SPDEs driven by space-time white noise, and with second order PDE operators, is limited to the case of a one-dimensional space variable. Also, there is no completely satisfactory theory of fully nonlinear SPDEs, see the work of Lions and Souganidis on viscosity solutions of SPDEs [15].

New powerful methods have been introduced recently to deal with singular SPDEs, namely the theory of regularity structures due to M. Hairer [10], and the notion of paracontrolled distributions introduced by Gubinelli, Imkeller and Perkowski [8]. We shall not discuss those approaches in the present notes.

1.2 Motivation

We now introduce several models from various fields, which are expressed as SPDEs.

1.2.1 Turbulence

Several mathematicians and physicists have advocated that the Navier–Stokes equation with additive white noise forcing is a suitable model for turbulence. This equation in dimension $d = 2$ or 3 reads

$$\begin{cases} \dfrac{\partial u}{\partial t}(t,x) = \nu \Delta u(t,x) + \displaystyle\sum_{i=1}^{d} u_i(t,x)\dfrac{\partial u}{\partial x_i}(t,x) + \dfrac{\partial W}{\partial t}(t,x) \\ u(0,x) = u_0(x), \end{cases}$$

where $u(t,x) = (u_1(t,x), \ldots, u_d(t,x))$ is the velocity of the fluid at time t and point x. The noise term is often chosen of the form

$$W(t,x) = \sum_{k=1}^{\ell} W^k(t)e_k(x),$$

where $\{W^1(t), \ldots, W^\ell(t), \ t \geq 0\}$ are mutually independent standard Brownian motions.

1.2.2 Population dynamics, population genetics

The following model was proposed by D. Dawson in 1972 for the evolution of the density of a population -

$$\frac{\partial u}{\partial t}(t,x) = v\frac{\partial^2 u}{\partial x^2}(t,x) + \alpha\sqrt{u}(t,x)\mathring{W}(t,x),$$

where $\mathring{W}$ is a space-time white noise. In this case, one can derive closed equations for the first two moments

$$m(t,x) = \mathbb{E}[u(t,x)], \quad V(t,x,y) = \mathbb{E}[u(t,x)u(t,y)].$$

One can approach this SPDE by a model in discrete space as follows. $u(t,i), i \in \mathbb{Z}$ denotes the number of individuals in the colony i at time t. Then

- $\frac{\alpha^2}{2}u(t,i)$ is both the birth and the death rate;
- $vu(t,i)$ is the migration rate, both from i to $i - 1$ and to $i + 1$.

W. Fleming has proposed an analogous model in population genetics, where the term $\alpha\sqrt{u}$ is replaced by $\alpha\sqrt{u(1-u)}$.

1.2.3 Neurophysiology

The following model has been proposed by J. Walsh [30], in order to describe the propagation of an electric potential in a neuron (which is identified with the interval $[0, L]$).

$$\frac{\partial V}{\partial t}(t,x) = \frac{\partial^2 V}{\partial x^2}(t,x) - V(t,x) + g(V(t,x))\mathring{W}(t,x).$$

Here again $\mathring{W}(t,x)$ denotes a space-time white noise.

1.2.4 Evolution of the curve of interest rate

This model was studied by R. Cont in 1998. Let $u(t,x)$, $0 \le x \le L$, $t \ge 0$ be the interest rate for a loan at time t, and duration x. We let

$$u(t,x) = r(t) + s(t)(Y(x) + X(t,x)),$$

where $Y(0) = 0$, $Y(L) = 1$; $X(t,0) = 0$, $X(t,L) = 1$; $\{(r(t), s(t)),\ t \ge 0\}$ is a two-dimensional diffusion process, and X solves the following parabolic SPDE

$$\frac{\partial X}{\partial t}(t,x) = \frac{k}{2}\frac{\partial^2 X}{\partial x^2}(t,x) + \frac{\partial X}{\partial x}(t,x) + \sigma(t, X(t,x))\mathring{W}(t,x).$$

Several authors have proposed a first-order parabolic SPDE (i.e. the above equation for X with $k = 0$), with a finite-dimensional noise.

1.2.5 Nonlinear filtering

Suppose a $\mathbb{R}^{d+k}$-valued process $\{(X_t, Y_t)\, t \ge 0\}$ is a solution of the system of SDEs

$$
\begin{cases}
X_t = X_0 + \displaystyle\int_0^t b(s, X_s, Y)ds + \int_0^t f(s, X_s, Y)dV_s + \int_0^t g(s, X_s, Y)dW_s \\
Y_t = \displaystyle\int_0^t h(s, X_s, Y)ds + W_t,
\end{cases}
$$

where the coefficients b, f, g and h may depend at each time s upon the whole past of Y before time s. We are interested in the evolution in t of the conditional law of X_t, given $\mathcal{F}_t^Y = \sigma\{Y_s,\ 0 \le s \le t\}$. It is known that if we denote by $\{\sigma_t,\ t \ge 0\}$ the measure-valued process solution of the Zakai equation

$$\sigma_t(\varphi) = \sigma_0(\varphi) + \int_0^t \sigma_s(L_{sY}\varphi)ds + \sum_{\ell=1}^k \int_0^t \sigma_s(L_{sY}^\ell \varphi)dY_s^\ell,\quad t \ge 0, \varphi \in C_b^\infty(\mathbb{R}^d),$$

where σ_0 denotes the law of X_0, and, if $a = ff^* + gg^*$,

$$L_{sY}\varphi(x) = \frac{1}{2}\sum_{i,j=1}^d a_{ij}(t,x,Y)\frac{\partial^2 \varphi}{\partial x_i \partial x_j}(x) + \sum_{i=1}^d b_i(t,x,Y)\frac{\partial \varphi}{\partial x_i}(x),$$

$$L_{sY}^\ell \varphi(x) = h_\ell(t,x,Y)\varphi(x) + \sum_{j=1}^d g_{i\ell}(t,x,Y)\frac{\partial \varphi}{\partial x_i}(x),$$

then

$$\mathbb{E}(\varphi(X_t)|\mathcal{F}_t) = \frac{\sigma_t(\varphi)}{\sigma_t(1)},$$

i.e. σ_t, is equal, up to a normalization factor, to the conditional law of X_t, given $\mathcal{F}_t$, see e.g. [25]. Note that whenever the random measure σ_t possesses a density $p(t,x)$, the latter satisfies the following SPDE

$$dp(t,x) = \left(\frac{1}{2}\sum_{i,j}\frac{\partial^2(a_{ij}p)}{\partial x_i \partial x_j}(t,x,Y)dt - \sum_i \frac{\partial(b_ip)}{\partial x_i}(t,x,Y)\right)dt$$
$$+ \sum_\ell \left(h_\ell p(t,x,Y) - \sum_i \frac{\partial(g_{i\ell}p)}{\partial x_i}(t,x,Y)\right)dY_t^\ell.$$

1.2.6 Movement by mean curvature in a random environment

Suppose that each point of a hypersurface in $\mathbb{R}^d$ moves in the direction normal to the hypersurface, with a speed given by

$$dV(x) = v_1(Du(x), u(x))dt + v_2(u(x)) \circ dW_t,$$

where $\{W_t,\ t \geq 0\}$ is a one-dimensional standard Brownian motion, and the notation $\circ$ means that the stochastic integral is understood in the Stratonovich sense.

The hypersurface at time t is a level set of the function $\{u(t,x),\ x \in \mathbb{R}^d\}$, where u solves a nonlinear SPDE of the form

$$du(t,x) = F(D^2u, Du)(t,x)dt + H(Du)(t,x) \circ dW_t,$$

where

$$F(X, p) = \text{tr}\left[\left(I - \frac{p \otimes p}{|p|^2}\right)X\right], \quad H(p) = \alpha|p|.$$

1.2.7 Hydrodynamic limit of particle systems

The following model was proposed by L. Bertini and G. Giacomin [2]. The idea is to describe the movement of a curve in $\mathbb{R}^2$ which is the interface between two substances, such as water and ice. The true model should be in $\mathbb{R}^3$, but this is an interesting simplified model.

Consider first a discrete model, where the interface belongs to the set

$$\Lambda = \{\xi \in \mathbb{Z}^\mathbb{Z},\ |\xi(x+1) - \xi(x)| = 1,\ \forall x \in \mathbb{Z}\}.$$

We describe the infinitesimal generator of the process of interest as follows. For any $\varepsilon > 0$, we define the infinitesimal generator

$$L_\varepsilon(\xi) = \sum_{x \in \mathbb{Z}} \left[c_\varepsilon^+(x, \xi)\{f(\xi + 2\delta_x) - f(\xi)\} \right.$$
$$\left. + c_\varepsilon^-(x, \xi)\{f(\xi - 2\delta_x) - f(\xi)\} \right],$$

where

$$\delta_x(y) = \begin{cases} 0, & \text{if } y \neq x; \\ 1, & \text{if } y = x; \end{cases}$$

$$c_\varepsilon^+(x, \xi) = \begin{cases} \frac{1}{2} + \sqrt{\varepsilon}, & \text{if } \xi(x) = \frac{\xi(x-1)+\xi(x+1)}{2} - 1; \\ 0, & \text{if not}; \end{cases}$$

$$c_\varepsilon^-(x, \xi) = \begin{cases} \frac{1}{2}, & \text{if } \xi(x) = \frac{\xi(x-1)+\xi(x+1)}{2} + 1; \\ 0, & \text{if not}. \end{cases}$$

Define $\{\xi_t^\varepsilon,\ t \geq 0\}$ as the jump Markov process with generator L^ε, and

$$u_\varepsilon(t, x) = \sqrt{\varepsilon}\left(\xi_{t/\varepsilon^2}\left(\frac{x}{\varepsilon}\right) - \left(\frac{1}{2\varepsilon^{3/2}} - \frac{1}{24\varepsilon^{1/2}}\right)t\right),$$

then we have the following result

Theorem 1.1. *If* $\sqrt{\varepsilon}\xi_0^\varepsilon\left(\frac{x}{\varepsilon}\right) \Rightarrow u_0(x)$, *and some technical conditions are met, then* $u_\varepsilon(t, x) \Rightarrow u(t, x)$, *where u solves (at least formally) the following SPDE*

$$\begin{cases} \dfrac{\partial u}{\partial t}(t, x) = \dfrac{1}{2}\dfrac{\partial^2 u}{\partial x^2}(t, x) - \dfrac{1}{2}\left|\dfrac{\partial u}{\partial x}(t, x)\right|^2 + \mathring{W}(t, x), \\ u(0, x) = u_0(x), \end{cases}$$

where $\mathring{W}$ denotes the space-time white noise.

The last SPDE is called the KPZ equation, after Kardar, Parisi and Zhang. Note that if we define $v(t, x) = \exp[-u(t, x)]$, we have the following equation for v

$$\frac{\partial v}{\partial t}(t, x) = \frac{1}{2}\frac{\partial^2 v}{\partial x^2}(t, x) - v(t, x)\mathring{W}(t, x).$$

If we regularize $\mathring{W}$ in space, then we construct corresponding sequences v_n and u_n, which satisfy

$$\frac{\partial v_n}{\partial t}(t, x) = \frac{1}{2}\frac{\partial^2 v_n}{\partial x^2}(t, x) - v_n(t, x)\,\mathring{W}_n(t, x),$$

and

$$\frac{\partial u_n}{\partial t}(t, x) = \frac{1}{2}\frac{\partial^2 u_n}{\partial x^2}(t, x) - \frac{1}{2}\left(\left|\frac{\partial u_n}{\partial x}(t, x)\right|^2 - c_n\right) + \mathring{W}_n(t, x),$$

where $c_n \to 0$, as $n \to \infty$.

1.2.8 Fluctuations of an interface on a wall

Funaki and Olla [7] proposed the following model in discrete space for the fluctuations of the microscopic height of an interface on a wall (the interface is forced to stay above the wall)

$$
\begin{cases}
dv_N(t,x) = - \left[V'(v_N(t,x) - v_N(t,x-1)) + V'(v_N(t,x) - v_N(t,x+1)) \right] dt \\
\qquad\qquad + dW(t,x) + dL(t,x), \quad t \geq 0, x \in \Gamma = \{1, 2, \ldots, N-1\}, \\
v_N(t,x) \geq 0, \quad L(t,x) \text{ is nondecreasing in } t, \text{ for all } x \in \Gamma \\
\displaystyle\int_0^\infty v_N(t,x) dL(t,x) = 0, \text{ for all } x \in \Gamma \\
v_N(t,0) = v_N(t,N) = 0, \quad t \geq 0,
\end{cases}
$$

where $V \in C^2(\mathbb{R})$ is symmetric and V'' is positive, bounded and bounded away from zero, and $\{W(t,1), \ldots, W(t, N-1), t \geq 0\}$ are mutually independent standard Brownian motions. The above is a coupled system of reflected SDEs. Assuming that $v_N(0, \cdot)$ is a random vector whose law is the invariant distribution of the solution of that system of reflected SDEs, one considers the rescaled macroscopic height

$$
\bar{v}_N(t,x) = \frac{1}{N} \sum_{y \in \Gamma} v_N(N^2 t, y) \mathbf{1}_{[y/N - 1/2N, y/N + 1/2N]}(x), \quad 0 \leq x \leq 1,
$$

which here converges to 0, as $N \to \infty$. Now the fluctuations, defined by

$$
u_N(t,x) = \frac{1}{\sqrt{N}} \sum_{y \in \Gamma} v_N(N^2 t, y) \mathbf{1}_{[y/N - 1/2N, y/N + 1/2N]}(x), \quad 0 \leq x \leq 1,
$$

converge, as $N \to \infty$, towards the solution of the reflected stochastic heat equation

$$
\begin{cases}
\dfrac{\partial u}{\partial t}(t,x) = v \dfrac{\partial^2 u}{\partial x^2}(t,x) + \mathring{W}(t,x) + \xi(t,x) \\
u(t,x) \geq 0, \ \xi \text{ is a random measure}, \ \displaystyle\int_{\mathbb{R}_+ \times [0,1]} u(t,x)\xi(dt, dx) = 0 \\
u(t,0) = u(t,1) = 0,
\end{cases}
$$

where $\mathring{W}(t,x)$ stands for the "space-time" white noise, and v is a constant which is in particular a function of V. Note that this reflected stochastic heat equation has been studied in Nualart and Pardoux [22], and will be discussed below in Section 3.8.

Chapter 2
SPDEs as Infinite-Dimensional SDEs

2.1 Introduction

The aim of this chapter is to describe by now classical results concerning mostly linear and semilinear SPDEs, considered as SDEs in a Hilbert or Banach space. We start with a short introduction to the Itô calculus in Hilbert space. We then briefly present the semigroup approach to linear SPDEs. We refer the reader to the monograph of Da Prato and Zabczyk [4] for a complete treatment of this approach for linear and semilinear SPDEs.

The main topic of this chapter is the presentation of the variational approach to SPDEs. We start with the case of deterministic PDEs, then present the theory of monotone-coercive SPDEs. This theory was first developed by the author, see [23], [24] and [25], and further improved by Krylov and Rozovski, see in particular [13] and [28]. We next present the extension to SPDEs of the compactness method, which is the second method developed by J.L. Lions [14] for the study of nonlinear PDEs, and at the same time constitutes an extension to SPDEs of the martingale approach problem to SDEs, due to Stroock and Varadhan [29]. We do this by presenting the theory developed in the unfortunately unpublished thesis of M. Viot.

2.2 Itô Calculus in Hilbert space

Let $(\Omega, \mathcal{F}, (\mathcal{F}_t), \mathbb{P})$ be a probability space equipped with a filtration $(\mathcal{F}_t)$ which is assumed to be right continuous and such that $\mathcal{F}_0$ contains all the $\mathbb{P}$-null sets of $\mathcal{F}$. A stochastic process $X : \Omega \times \mathbb{R}_+ \mapsto \mathbb{X}$ (where $\mathbb{X}$ can be, for example, $\mathbb{R}^d$, a Hilbert or a Banach space) is said to be *progressively measurable* if for any $t > 0$, the mapping $(\omega, s) \mapsto X(\omega, s)$ from $\Omega \times [0, t]$ into $\mathbb{X}$ is $(\mathcal{F}_t \otimes \mathcal{B}_{[0,t]}, \mathcal{B}_{\mathbb{X}})$-measurable. Here we denote by $\mathcal{B}_A$ the Borel σ-algebra of subsets of A.

© The Author(s), under exclusive license to Springer Nature Switzerland AG 2021
É. Pardoux, *Stochastic Partial Differential Equations*,
SpringerBriefs in Mathematics, https://doi.org/10.1007/978-3-030-89003-2_2

Martingales

Let H be a Hilbert space, and $\{M_t,\ 0 \leq t \leq T\}$ be a continuous H-valued martingale, which is such that $\sup_{0 \leq t \leq T} \mathbb{E}(\|M_t\|^2) < \infty$.

Then $\{\|M_t\|^2,\ 0 \leq t \leq T\}$ is a continuous real-valued submartingale, and there exists a unique continuous increasing $\mathcal{F}_t$-adapted process $\{\langle M \rangle_t,\ 0 \leq t \leq T\}$ such that $\{\|M_t\|^2 - \langle M \rangle_t,\ 0 \leq t \leq T\}$ is a martingale.

We denote by $\{M_t \otimes M_t,\ 0 \leq t \leq T\}$ the $\mathcal{L}_+^1(H)$-valued process defined by

$$((M_t \otimes M_t)h, k)_H = (M_t, h)_H \times (M_t, h)_H,$$

$h, k \in H$. We have used the notation $\mathcal{L}_+^1(H)$ to denote the set of self-adjoint, semi-definite, linear, positive trace-class operators from H into itself. We have the following theorem, whose last assertion is due to Métivier and Pistone, see Métivier [17].

Theorem 2.1. *To any continuous square integrable H-valued martingale $\{M_t,\ 0 \leq t \leq T\}$, we can associate a unique continuous adapted increasing $\mathcal{L}_+^1(H)$-valued process $\{\langle\langle M \rangle\rangle_t,\ 0 \leq t \leq T\}$ such that $\{M_t \otimes M_t - \langle\langle M \rangle\rangle_t,\ 0 \leq t \leq T\}$ is a martingale. Moreover, there exists a unique predictable $\mathcal{L}_+^1(H)$-valued process $\{Q_t, 0 \leq t \leq T\}$ such that*

$$\langle\langle M \rangle\rangle_t = \int_0^t Q_s \mathrm{d}\langle M \rangle_s, \quad 0 \leq t \leq T.$$

Note that since Tr is a linear operator,

$$\mathrm{Tr}(M_t \otimes M_t - \langle\langle M \rangle\rangle_t) = \|M_t\|^2 - \mathrm{Tr}\langle\langle M \rangle\rangle_t$$

is a real-valued martingale, hence $\mathrm{Tr}\langle\langle M \rangle\rangle_t = \langle M \rangle_t$. Consequently, we have that $\langle M \rangle_t = \int_0^t \mathrm{Tr}Q_s \mathrm{d}\langle M \rangle_s$, and

$$\mathrm{Tr}Q_t = 1, \quad t \text{ a.e., a.s.} \tag{2.1}$$

Example 2.2 (H-valued Wiener process). Let $\{B_t^k,\ t \geq 0,\ k \in \mathbb{N}\}$ be a collection of mutually independent standard scalar Brownian motions, and $Q \in \mathcal{L}_+^1(H)$. If $\{e_k,\ k \in \mathbb{N}\}$ is an orthonormal basis of H, then the process

$$W_t = \sum_{k \in \mathbb{N}} B_t^k Q^{1/2} e_k, \quad t \geq 0$$

is an H-valued square integrable martingale, with $\langle W \rangle_t = \mathrm{Tr}Q \times t$. It is called an H-valued Wiener process, or Brownian motion.

Conversely, if $\{M_t,\ 0 \leq t \leq T\}$ is a continuous H-valued martingale such that $\langle M \rangle_t = c \times t$ and $Q_t = Q$, where $c \in \mathbb{R}$ and $Q \in \mathcal{L}_+^1(H)$ are deterministic, then $\{M_t,\ 0 \leq t \leq T\}$ is an H-valued Wiener process (this is an infinite-dimensional version of a well-known theorem due to P. Lévy).

Example 2.3 (Cylindrical Brownian motion). This should be called a "counter-example", rather than an example. Let again $\{B_t^k,\ t \geq 0,\ k \in \mathbb{N}\}$ be a collection of mutually independent standard scalar Brownian motions, and $\{e_k,\ k \in \mathbb{N}\}$ an orthonormal basis of H. Then the series

$$W_t = \sum_{k \in \mathbb{N}} B_t^k e_k$$

does not converge in H. In fact it converges in any larger space K such that the injection from H into K is Hilbert–Schmidt. We shall call such a process a cylindrical Wiener process on H (which does not take its values in H!). Formally, $\langle\langle W \rangle\rangle_t = tI$, which is not trace class!

Stochastic integral

Let $\{\varphi_t,\ 0 \leq t \leq T\}$ be a predictable H-valued process such that

$$\int_0^T (Q_t\varphi_t, \varphi_t)_H d\langle M \rangle_t < \infty \quad \text{a.s.}$$

Then we can define the stochastic integral

$$\int_0^t (\varphi_s, dM_s)_H, \quad 0 \leq t \leq T.$$

More precisely, we have that

$$\int_0^t (\varphi_s, dM_s)_H = \lim_{n \to \infty} \sum_{i=1}^{n-1} \left(\frac{1}{t_i^n - t_{i-1}^n} \int_{t_{i-1}^n}^{t_i^n} \varphi_s ds, M_{t_{i+1}^n \wedge t} - M_{t_i^n \wedge t} \right)_H,$$

with for example $t_i^n = iT/n$. The above limit holds in probability.

The process $\{\int_0^t (\varphi_s, dM_s)_H,\ 0 \leq t \leq T\}$ is a continuous $\mathbb{R}$-valued local martingale, with

$$\left\langle \int_0^{\cdot} (\varphi_s, dM_s)_H \right\rangle_t = \int_0^t (Q_s\varphi_s, \varphi_s)_H d\langle M \rangle_s,$$

and if, moreover,

$$\mathbb{E} \int_0^T (Q_t\varphi_t, \varphi_t)_H d\langle M \rangle_t < \infty,$$

then the above stochastic integral is a square integrable martingale.

Stochastic integral with respect to a cylindrical Brownian motion

Let again $\{\varphi_t,\ 0 \leq t \leq T\}$ be a progressively measurable H-valued process. We suppose now that

$$\int_0^T \|\varphi_t\|_H^2 dt < \infty \quad \text{a.s.}$$

It is then not very difficult to show that

$$\int_0^t (\varphi_s, dW_s) = \lim_{n \to \infty} \sum_{k=1}^n \int_0^t (\varphi_s, e_k) dB_s^k$$

exists as a limit in probability.

Itô formula

Let $\{X_t\}$, $\{V_t\}$ and $\{M_t\}$ be progressively measurable H-valued processes, where

- $X_t = X_0 + V_t + M_t, \quad t \geq 0,$
- $\{V_t\}$ is a bounded variation process with $V_0 = 0,$
- $\{M_t\}$ is a local martingale with $M_0 = 0.$

Let $\Phi : H \to \mathbb{R}$ be such that $\Phi \in C^1(H; \mathbb{R})$, and for any $h \in H$, $\Phi''(h)$ exists in the Gateaux sense, and moreover $\forall Q \in \mathcal{L}^1(H)$, the mapping $h \to \text{Tr}(\Phi''(h)Q)$ is continuous. Then we have

$$\Phi(X_t) = \Phi(X_0) + \int_0^t (\Phi'(X_s), dV_s) + \int_0^t (\Phi'(X_s), dM_s)$$

$$+ \frac{1}{2} \int_0^t \text{Tr}(\Phi''(X_s)Q_s)d\langle M\rangle_s.$$

Example 2.4. The case where $\Phi(h) = \|h\|_H^2$ will be important in what follows. In that case, we have

$$\|X_t\|^2 = \|X_0\|^2 + 2\int_0^t (X_s, dV_s) + 2\int_0^t (X_s, dM_s) + \langle M\rangle_t,$$

since here $\Phi''/2 = I$, and $\text{Tr}Q_s = 1$, see (2.1).

2.3 SPDEs with Additive Noise

This is the simplest case, where the existence-uniqueness theory needs almost no more than the theory of deterministic PDEs. We are motivated by the following two examples:

1. The heat equation with additive noise. Let us consider our last example from section 1.2.8, but without the reflection, i.e. the SPDE (here in arbitrary dimension, $x \in D \subset \mathbb{R}^d$)

$$\begin{cases} \dfrac{\partial u}{\partial t}(t,x) = \nu \Delta u(t,x) + \dfrac{\partial W}{\partial t}(t,x), & t \geq 0,\ x \in D \\ u(0,x) = u_0(x), & u(t,x) = 0,\ t \geq 0,\ x \in \partial D, \end{cases}$$

where $\{W(t,x),\ t \geq 0, x \in D\}$ denotes a Wiener process with respect to the time variable, with arbitrary correlation in the spatial variable (possibly white in space).

2. The two-dimensional Navier–Stokes equation with additive finite-dimensional noise. Its vorticity formulation is as follows

$$\begin{cases} \dfrac{\partial \omega}{\partial t}(t,x) + B(\omega,\omega)(t,x) = \nu \Delta \omega(t,x) + \dfrac{\partial W}{\partial t}(t,x) \\ \omega(0,x) = \omega_0(x), \end{cases}$$

where $x = (x_1,x_2) \in \mathbf{T}^2$, the two-dimensional torus $[0,2\pi] \times [0,2\pi]$, $\nu > 0$ is the viscosity constant, $\frac{\partial W}{\partial t}$ is a white-in-time stochastic forcing of the form

$$W(t,x) = \sum_{k=1}^{\ell} W_k(t)e_k(x),$$

where $\{W_1(t),\ldots,W_\ell(t)\}$ are mutually independent standard Brownian motions, and

$$B(\omega,\tilde{\omega}) = \sum_{i=1}^{2} u_i(x)\frac{\partial \tilde{\omega}}{\partial x_i}(x),$$

where $u = \mathcal{K}(\omega)$. Here $\mathcal{K}$ is the Biot–Savart law, which in the two-dimensional periodic setting can be expressed as

$$\mathcal{K}(\omega) = \sum_{k \in \mathbb{Z}_*^2} \frac{k^\perp}{|k|^2} \big[\beta_k \cos(k \cdot x) - \alpha_k \sin(k \cdot x)\big], \tag{2.2}$$

where $k^\perp = (-k_2,k_1)$ and $\omega(t,x) = \sum_{k \in \mathbb{Z}_*^2} \alpha_k \cos(k \cdot x) + \beta_k \sin(k \cdot x)$ with $\mathbb{Z}_*^2 = \{(j_1,j_2) \in \mathbb{Z}^2 : j_2 \geq 0, |j| > 0\}$.

Let us start with some results on PDEs, sketching two different approaches.

2.3.1 The semigroup approach to linear parabolic PDEs

First consider the following abstract linear parabolic equation

$$\begin{cases} \dfrac{\partial u}{\partial t}(t) = Au(t),\ t \geq 0 \\ u(0) = u_0, \end{cases}$$

where A is a (possibly unbounded) linear operator on some Hilbert space H, i.e. A maps its domain $D(A) \subset H$ into H. Suppose that $u_0 \in H$, and we are looking for a solution which should take its values in H. For each $t > 0$, the mapping $u_0 \to u(t)$ is a linear mapping $P(t) \in \mathcal{L}(H)$, and the mappings $\{P(t), \ t \geq 0\}$ form a semigroup, in the sense that $P(t + s) = P(t)P(s)$. A is called the infinitesimal generator of this semigroup. Suppose now that $H = L^2(D)$, where D is some domain in $\mathbb{R}^d$. Then the linear operator $P(t)$ has a kernel $p(t, x, y)$ such that $\forall h \in L^2(D)$,

$$[P(t)h](x) = \int_D p(t, x, y)h(y)dy.$$

Example 2.5. If $D = \mathbb{R}^d$, and $A = \frac{1}{2}\Delta$, then

$$p(t, x, y) = \frac{1}{(2\pi t)^{d/2}} \exp\left(-\frac{|x - y|^2}{2t}\right).$$

Consider now the PDE

$$\begin{cases} \dfrac{\partial u}{\partial t}(t) = Au(t) + f(t), \ t \geq 0 \\ u(0) = u_0, \end{cases}$$

where $f(\cdot)$ is an H-valued function of t. The solution of this last equation is given by the *variation of constants formula*

$$u(t) = P(t)u_0 + \int_0^t P(t - s)f(s)ds.$$

Consider now the parabolic equation with additive white noise, i.e.

$$\begin{cases} \dfrac{du}{dt}(t) = Au(t) + \dfrac{dW}{dt}(t), \ t \geq 0 \\ u(0) = u_0, \end{cases} \tag{2.3}$$

where $\{W(t), \ t \geq 0\}$ is an H-valued Wiener process. Then the variation of constants formula, generalized to this situation, yields the following formula for $u(t)$:

$$u(t) = P(t)u_0 + \int_0^t P(t - s)dW(s),$$

in terms of a Wiener integral (a Wiener integral is an Itô integral whose integrand is deterministic). In the case $H = L^2(D)$, $W(t) = W(t, x)$ and this formula can be rewritten more explicitly as follows:

$$u(t, x) = \int_D p(t, x, y)u_0(y)dy + \int_0^t \int_D p(t - s, x, y)W(ds, y)dy.$$

In the case of the cylindrical Wiener process, i.e. if the equation is driven by space-time white noise, then the above formula takes the form

$$u(t,x) = \int_D p(t,x,y)u_0(y)\mathrm{d}y + \int_0^t \int_D p(t-s,x,y)W(\mathrm{d}s,\mathrm{d}y),$$

where $\{W(t,x),\ t \ge 0, x \in D\}$ denotes the so-called Brownian sheet, and the above is a two-parameter stochastic integral, which we will discuss in more detail in chapter 3.

We just considered a case where $W(t)$ does not take its values in H. Let us now discuss the opposite case, where $W(t)$ takes its values not only in H, but in fact in $D(A)$. Then considering again the equation (2.3), and defining $v(t) = u(t) - W(t)$, we have the following equation for v:

$$\begin{cases} \dfrac{\mathrm{d}v}{\mathrm{d}t}(t) = Av(t) + AW(t) \\ \\ v(0) = u_0, \end{cases}$$

which can be solved ω by ω, without any stochastic integration.

2.3.2 *The variational approach to linear and nonlinear parabolic PDEs*

We now sketch the variational approach to deterministic PDEs, which was developed among others by J.L. Lions. We first consider the case of linear equations.

Linear equations

From now on, A will denote an extension of the unbounded operator from the previous section. That is, instead of considering

$$A\ :\ D(A) \longrightarrow H,$$

we shall consider

$$A\ :\ V \longrightarrow V',$$

where

$$D(A) \subset V \subset H \subset V'.$$

More precisely, the framework is as follows.

H is a separable Hilbert space. We shall denote by $|\cdot|_H$ or simply by $|\cdot|$ the norm in H and by $(\cdot,\cdot)_H$ or simply $(\cdot,\cdot)$ its scalar product. Let $V \subset H$ be a reflexive Banach space, which is dense in H, with continuous injection. We shall denote by $\|\cdot\|$ the norm in V. We shall identify H with its dual H', and consider H' as a subspace of the dual V' of V, again with continuous injection. We then have the situation

$$V \subset H \simeq H' \subset V'.$$

More precisely, we assume that the duality pairing $\langle \cdot, \cdot \rangle$ between V and V' is such that whenever $u \in V$ and $v \in H \subset V'$, $\langle u, v \rangle = (u, v)_H$. Finally, we shall denote by $\| \cdot \|_*$ the norm in V', defined by

$$\|v\|_* = \sup_{u \in V, \ \|u\| \le 1} \langle u, v \rangle.$$

We can without loss of generality assume that whenever $u \in V$, $|u| \le \|u\|$. It then follows (exercise) that if again $u \in V$, $\|u\|_* \le |u| \le \|u\|$.

Now suppose an operator $A \in \mathcal{L}(V, V')$ is given, which is assumed to satisfy the following *coercivity* assumption:

$$\begin{cases} \exists \lambda, \alpha > 0 \text{ such that } \forall u \in V, \\ 2\langle Au, u \rangle + \alpha \|u\|^2 \le \lambda |u|^2. \end{cases}$$

Example 2.6. Let D be an open domain in $\mathbb{R}^d$. We let $H = L^2(D)$ and $V = H^1(D)$, where

$$H^1(D) = \{ u \in L^2(D); \ \frac{\partial u}{\partial x_i} \in L^2(D), \ i = 1, \ldots, d \}.$$

Equipped with the scalar product

$$((u, v)) = \int_D u(x) v(x) dx + \sum_{i=1}^d \int_D \frac{\partial u}{\partial x_i}(x) \frac{\partial v}{\partial x_i}(x) dx,$$

$H^1(D)$ is a Hilbert space, as is $H_0^1(D)$, which is the closure in $H^1(D)$ of the set $C_K^\infty(D)$ of smooth functions with support in a compact subset of D. We now let

$$\Delta = \sum_{i=1}^d \frac{\partial^2}{\partial x_i^2}.$$

We have $\Delta \in \mathcal{L}(H^1(D), [H^1(D)]')$, and also $\Delta \in \mathcal{L}(H_0^1(D), [H_0^1(D)]')$. Note that, provided the boundary ∂D of D is sufficiently smooth, $H_0^1(D)$ can be identified with the closed subset of $H^1(D)$ consisting of those functions which are zero on the boundary ∂D (one can indeed make sense of the trace of $u \in H^1(D)$ on the boundary ∂D). We have $[H_0^1(D)]' = H^{-1}(D)$, where any element of $H^{-1}(D)$ can be put in the form

$$f + \sum_{i=1}^d \frac{\partial g_i}{\partial x_i},$$

where $f, g_1, \ldots, g_d \in L^2(D)$.

Consider the linear parabolic equation

$$\begin{cases} \frac{du}{dt}(t) = Au(t) + f(t), \ t \ge 0; \\ u(0) = u_0. \end{cases} \tag{2.4}$$

The following theorem holds.

Theorem 2.7. *If $A \in \mathcal{L}(V, V')$ is coercive, $u_0 \in H$ and $f \in L^2(0,T;V')$, then equation (2.4) has a unique solution $u \in L^2(0,T;V)$, which also belongs to $C([0,T];H)$.*

We first need to show the following interpolation result, which is Lemma 2.14 below in the particular case $M \equiv 0$.

Lemma 2.8. *If $u \in L^2(0,T;V)$, $t \to u(t)$ is absolutely continuous with values in V', $\frac{du}{dt} \in L^2(0,T;V')$ and $u(0) \in H$, then $u \in C([0,T];H)$ and*

$$\frac{d}{dt}|u(t)|^2 = 2\left\langle \frac{du}{dt}(t), u(t) \right\rangle, \ t \ a.e.$$

PROOF OF THEOREM 2.7

Uniqueness Let $u, v \in L^2(0,T;V)$ be two solutions of equation (2.4). Then the difference $u - v$ solves

$$\frac{d(u-v)}{dt}(t) = A(u(t) - v(t)),$$

$$u(0) - v(0) = 0.$$

Then from Lemma 2.8,

$$|u(t) - v(t)|^2 = 2\int_0^t \langle A(u(s) - v(s)), u(s) - v(s)\rangle ds \le \lambda \int_0^t |u(s) - v(s)|^2 ds,$$

and Gronwall's lemma implies that $u(t) - v(t) = 0$, $\forall t \ge 0$.

Existence We use a Galerkin approximation. Let $\{e_k, \ k \ge 1\}$ denote an orthonormal basis of H, made of elements of V. For each $n \ge 1$, we define

$$V_n = \text{span}\{e_1, e_2, \ldots, e_n\}.$$

For all $n \ge 1$, there exists a function $u_n \in C([0,T];V_n)$ such that for all $1 \le k \le n$,

$$\frac{d}{dt}(u_n(t), e_k) = \langle Au_n(t), e_k \rangle + \langle f(t), e_k \rangle,$$

$$(u_n(0), e_k) = (u_0, e_k).$$

u_n is the solution of a finite-dimensional linear ODE. We now prove the following uniform estimate

$$\sup_n \left[\sup_{0 \le t \le T} |u_n(t)|^2 + \int_0^T \|u_n(t)\|^2 dt \right] < \infty. \tag{2.5}$$

It is easily seen that

$$|u_n(t)|^2 = \sum_{k=1}^n (u_0, e_k)^2 + 2\int_0^t \langle Au_n(s) + f(s), u_n(s)\rangle ds.$$

Hence we deduce from the coercivity of A that

$$|u_n(t)|^2 + \alpha \int_0^t \|u_n(s)\|^2 ds \le |u_0|^2 + \int_0^T \|f(s)\|_*^2 ds + (\lambda + 1) \int_0^t |u_n(s)|^2 ds,$$

and (2.5) follows from Gronwall's lemma.

Now there exists a subsequence, which, by an abuse of notation, we still denote by $\{u_n\}$, which converges in $L^2(0, T; V)$ weakly to some u. Since A is linear and continuous from V into V', it is also continuous with respect to the weak topologies, and taking the limit in the approximating equation, we have a solution of (2.4).   □

Let us now indicate how this approach can be extended to nonlinear equations.

Nonlinear equations

Suppose now that $A : V \to V'$ is a nonlinear operator satisfying again the coercivity assumption. We can repeat the first part of the above proof. However, taking the limit in the approximating sequence is now much more involved. The problem is the following. While a continuous linear operator is continuous with respect to the weak topologies, a nonlinear operator which is continuous for the strong topologies, typically fails to be continuous with respect to the weak topologies.

In the framework which has been exposed in this section, there are two possible solutions, which necessitate two different assumptions.

1. **Monotonicity.** If we assume that the nonlinear operator A satisfies in addition the condition

$$\langle A(u) - A(v), u - v \rangle \le \lambda |u - v|^2,$$

   together with some boundedness condition of the type $\|A(u)\|_* \le c(1 + \|u\|)$, and some continuity condition, then the above difficulty can be solved. Indeed, following the proof in the linear case, we show both that $\{u_n\}$ is a bounded sequence in $L^2(0, T; V)$ and that $\{A(u_n)\}$ is a bounded sequence in $L^2(0, T; V')$. Hence there exists a subsequence, still denoted the same way, along which $u_n \to u$ in $L^2(0, T; V)$ weakly, and $A(u_n) \to \xi$ weakly in $L^2(0, T; V')$. It remains to show that $\xi = A(u)$. Let us explain the argument, in the case where the monotonicity assumption is satisfied with $\lambda = 0$. Then we have that for all $v \in L^2(0, T; V)$,

$$\int_0^T \langle A(u_n(t)) - A(v(t)), u_n(t) - v(t) \rangle dt \le 0.$$

The above expression can be developed into four terms, three of which converge without any difficulty to the sought limit. The only difficulty is with the term

$$\int_0^T \langle A(u_n(t)), u_n(t) \rangle dt = \frac{1}{2}(|u_n(T)|^2 - \sum_{k=1}^n (u_0, e_k)^2) - \int_0^T \langle f(t), u_n(t) \rangle dt.$$

Two of the three terms of the right-hand side converge. The first one *does not*. However, it is not hard to show that the subsequence can be chosen in such a way that $u_n(T) \to u(T)$ in H weakly, and the mapping which to a vector in H associates the square of its norm is convex and strongly continuous, hence it is the upper envelope of linear continuous (hence also weakly continuous) mappings, hence it is l.s.c. with respect to the weak topology of H, so

$$\liminf_n |u_n(T)|^2 \geq |u(T)|^2,$$

and consequently we have that, again for all $v \in L^2(0, T; V)$,

$$\int_0^T \langle \xi(t) - A(v(t)), u(t) - v(t) \rangle dt \leq 0.$$

We now choose $v(t) = u(t) - \theta w(t)$, with $\theta > 0$, divide by θ, and let $\theta \to 0$, yielding

$$\int_0^T \langle \xi(t) - A(u(t)), w(t) \rangle dt \leq 0.$$

Since w is an arbitrary element of $L^2(0, T, ; V)$, the left-hand side must vanish, hence $\xi \equiv A(u)$.

Example 2.9. The simplest example of an operator which is monotone in the above sense is an operator of the form

$$A(u)(x) = \Delta u(x) + f(u(x)),$$

where $f : \mathbb{R} \to \mathbb{R}$ is the sum of a Lipschitz and a decreasing function.

2. **Compactness** We now assume that the injection from V into H is compact (in the example $V = H^1(D)$, $H = L^2(D)$, this implies that D is bounded). Note that in the preceding arguments, there was no serious difficulty in proving that the sequence $\{\frac{du_n}{dt}\}$ is bounded in $L^2(0, T; V')$. But one can show the following compactness Lemma (see Lions [14]):

Lemma 2.10. *Let the injection from V into H be compact. If a sequence $\{u_n\}$ is bounded in $L^2(0, T; V)$, while the sequence $\{\frac{du_n}{dt}\}$ is bounded in $L^2(0, T; V')$, then one can extract a subsequence of the sequence $\{u_n\}$, which converges strongly in $L^2(0, T; H)$.*

Let us explain how this lemma can be used in the case of the Navier–Stokes equation. The nonlinear term is the sum of terms of the form $u_i(t, x)\frac{\partial u}{\partial x_i}$, i.e. the product of a term which converges strongly with a term which converges weakly, i.e. one can take the limit in that product.

PDE with additive noise

Let us now consider the parabolic PDE

$$\begin{cases} \dfrac{du}{dt}(t) = A(u(t)) + f(t) + \dfrac{dW}{dt}(t), \ t \geq 0; \\ u(0) = u_0. \end{cases}$$

If we assume that the trajectories of the Wiener process $\{W(t)\}$ belong to $L^2(0,T;V)$, then we can define $v(t) = u(t) - W(t)$, and note that v solves the PDE with random coefficients

$$\begin{cases} \dfrac{dv}{dt}(t) = A(v(t) + W(t)) + f(t), \ t \geq 0; \\ u(0) = u_0, \end{cases}$$

which can again be solved ω by ω, without any stochastic integration. However, we want to treat equations driven by a noise which does not necessarily takes its values in V, and also may not be additive.

2.4 The Variational Approach to SPDEs

The framework is the same as in the last subsection.

2.4.1 Monotone-coercive SDPEs

Let $A : V \to V'$ and for each $k \geq 1$, $B_k : V \to H$, so that $B = (B_k, \ k \geq 1) : V \to \mathcal{H} = \ell^2(H)$.

We make the following four basic assumptions:

Coercivity

$$(H1) \begin{cases} \exists \alpha > 0, \lambda, \nu \ \text{such that} \ \forall u \in V, \\ 2\langle A(u), u \rangle + |B(u)|^2_{\mathcal{H}} + \alpha \|u\|^2 \leq \lambda |u|^2 + \nu. \end{cases}$$

Monotonicity

$$(H2) \begin{cases} \exists \lambda > 0 \ \text{such that} \ \forall u, v \in V, \\ 2\langle A(u) - A(v), u - v \rangle + |B(u) - B(v)|^2_{\mathcal{H}} \leq \lambda |u - v|^2. \end{cases}$$

Linear growth

$$(H3) \quad \exists c > 0 \ \text{such that} \ \|A(u)\|_* \leq c(1 + \|u\|), \ \forall u \in V.$$

Weak continuity

$$(H4) \begin{cases} \forall u, v, w \in V, \\ \text{the mapping} \ \lambda \to \langle A(u + \lambda v), w \rangle \ \text{is continuous from} \ \mathbb{R} \ \text{into} \ \mathbb{R}. \end{cases}$$

Note that

$$|B(u)|^2_{\mathcal{H}} = \sum_{k=1}^{\infty} |B_k(u)|^2, \quad |B(u) - B(v)|^2_{\mathcal{H}} = \sum_{k=1}^{\infty} |B_k(u) - B_k(v)|^2.$$

We want to study the equation

$$u(t) = u_0 + \int_0^t A(u(s))ds + \int_0^t B(u(s))dW_s$$

$$= u_0 + \int_0^t A(u(s))ds + \sum_{k=1}^\infty \int_0^t B_k(u(s))dW_s^k, \qquad (2.6)$$

where $u_0 \in H$, and $\{W_t = (W_t^k, \ k = 1,2,\ldots), \ t \geq 0\}$ is a sequence of mutually independent $\mathcal{F}_t$-standard scalar Brownian motions. We shall look for a solution u whose trajectories should satisfy $u \in L^2(0,T;V)$, for all $T > 0$. Hence $A(u(\cdot)) \in L^2(0,T;V')$, for all $T > 0$. In fact, the above equation can be considered as an equation in the space V', or equivalently we can write the equation in the so-called *weak form*

$$(u(t),v) = (u_0,v) + \int_0^t \langle A(u(s)),v\rangle ds + \int_0^t (B(u(s)),v)dW_s, \ \forall v \in V, t \geq 0, \quad (2.7)$$

where the stochastic integral term should be interpreted as

$$\int_0^t (B(u(s)),v)dW_s = \sum_{k=1}^\infty \int_0^t (B_k(u(s)),v)dW_s^k.$$

Remark 2.11. Since $|u| \leq \|u\|$, it follows from $(H1) + (H3)$ that for some constant c', $|B(u)|_\mathcal{H} \leq c'(1 + \|u\|)$.

We can w.l.o.g. assume that λ is the same in $(H1)$ and in $(H2)$. In fact it suffices to treat the case $\lambda = 0$, since $v = e^{-\lambda t/2}u$ solves the same equation, with A replaced by

$$e^{-\lambda t/2}A(e^{\lambda t/2}\cdot) - \frac{\lambda}{2}I$$

and B replaced by

$$e^{-\lambda t/2}B(e^{\lambda t/2}\cdot),$$

and in most cases of interest this new pair satisfies $(H1)$ and $(H2)$ with $\lambda = 0$.

Remark 2.12. We can replace in $(H1)$ $\|u\|^2$ by $\|u\|^p$, with $p > 2$, provided we replace $(H3)$ by

$$(H3)_p \quad \exists c > 0 \text{ such that } \|A(u)\|_* \leq c(1 + \|u\|^{p-1}), \ \forall u \in V.$$

This modified set of assumptions is well adapted for treating certain nonlinear equations, see the last example in the next subsection. Note that the operator A can be the sum of several A_i's with different associated p_i's.

We can now state the main result of this section.

Theorem 2.13. *Under the assumptions* $(H1), (H2), (H3)$ *and* $(H4)$, *if* $u_0 \in H$, *there exists a unique adapted process* $\{u(t), \ t \geq 0\}$ *whose trajectories belong a.s. for any* $T > 0$ *to the space* $L^2(0,T;V) \cap C([0,T];H)$, *which is a solution to equation* (2.6).

An essential tool for the proof of this theorem is the following ad hoc Itô formula:

Lemma 2.14. *Let* $u_0 \in H$, $\{u(t), 0 \le t \le T\}$ *and* $\{v(t), 0 \le t \le T\}$ *be adapted processes with trajectories in* $L^2(0,T;V)$ *and* $L^2(0,T;V')$ *respectively, and* $\{M_t, 0 \le t \le T\}$ *be a continuous H-valued local martingale, such that*

$$u(t) = u_0 + \int_0^t v(s)ds + M_t.$$

Then

(i) $u \in C([0,T];H)$ *a.s.*
(ii) *the following formula holds* $\forall 0 \le t \le T$ *and a.s.*

$$|u(t)|^2 = |u_0|^2 + 2\int_0^t \langle v(s), u(s) \rangle ds + 2\int_0^t (u(s), dM_s) + \langle M \rangle_t.$$

PROOF: **Proof of (ii)** Since V is dense in H, there exists an orthonormal basis $\{e_k, k \ge 1\}$ of H with each $e_k \in V$. For the sake of this proof, we shall assume that V is a Hilbert space, and that the above basis is also orthogonal in V. Although these assumptions are not always true, they hold in many interesting examples. The general proof is more involved than the one which follows, see the comments after the proof for references. We have, with the notation $M_t^k = (M_t, e_k)$,

$$|u(t)|^2 = \sum_k (u(t), e_k)^2$$

$$= \sum_k \left[(u_0, e_k)^2 + 2\int_0^t \langle v(s), e_k \rangle (e_k, u(s)) ds + 2\int_0^t (u(s), e_k) dM_s^k + \langle M^k \rangle_t \right]$$

$$= |u_0|^2 + 2\int_0^t \langle v(s), u(s) \rangle ds + 2\int_0^t (u(s), dM_s) + \langle M \rangle_t.$$

Proof of (i) It clearly follows from our assumptions that $u \in C([0,T];V')$ a.s. Moreover, from (ii), $t \to |u(t)|$ is a.s. continuous. It suffices to show that $t \to u(t)$ is continuous into H equipped with its weak topology, since whenever $u_n \to u$ in H weakly and $|u_n| \to |u|$, then $u_n \to u$ in H strongly (an easy exercise, exploiting the fact that H is a Hilbert space). Now, clearly $u \in L^\infty(0,T;H)$ a.s., again thanks to (ii). Now let $h \in H$ and a sequence $t_n \to t$, as $n \to \infty$, be arbitrary. All we have to show is that $(u(t_n), h) \to (u(t), h)$ a.s. Let $\{h_m, m \ge 1\} \subset V$ be such that $h_m \to h$ in H, as $m \to \infty$. Let us choose an arbitrary $\varepsilon > 0$, and m_0 large enough, such that

$$\sup_{0 \le t \le T} |u(t)| \times |h - h_m| \le \varepsilon/2, \quad m \ge m_0.$$

It follows that

$$|(u(t), h) - (u(t_n), h)| \le |(u(t), h - h_{m_0})| + |(u(t) - u(t_n), h_{m_0})|$$
$$+ |(u(t_n), h - h_{m_0})|$$
$$\le \|u(t) - u(t_n)\|_* \times \|h_{m_0}\| + \varepsilon,$$

hence

$$\limsup_n |(u(t), h) - (u(t_n), h)| \leq \varepsilon,$$

and the result follows from the fact that ε is arbitrary.   □

The above lemma is proved under the assumption that there exists an operator $A \in \mathcal{L}(V, V')$ satisfying (H1) above with $B = 0$ in Pardoux [23] and [24]. The result as stated above was proved by Krylov and Rozovsky, see [28]. A new proof has recently been given by Krylov [12].

We give a further result, which will be needed below. It is proved similarly as the preceding result, see e.g. [24].

Lemma 2.15. *Under the assumptions of Lemma 2.14, let* Φ *be a functions from H into $\mathbb{R}$ satisfying*

(i) *all assumptions from the Itô formula in Section 2.2,*
(ii) $\Phi'(u) \in V$ *whenever* $u \in V$,
(iii) *the mapping* $u \to \Phi'(u)$ *is continuous from V into V equipped with the weak topology, and*
(iv) *for some c and all* $u \in V$,

$$\|\Phi'(u)\| \leq c(1 + \|u\|).$$

Then we have the Itô formula

$$\Phi(X_t) = \Phi(X_0) + \int_0^t \langle v_s, \Phi'(X_s) \rangle \mathrm{d}s + \int_0^t (\Phi'(X_s), \mathrm{d}M_s)$$
$$+ \frac{1}{2} \int_0^t \mathrm{Tr}(\Phi''(X_s) Q_s) \mathrm{d}\langle M \rangle_s.$$

PROOF OF THEOREM 2.13

Uniqueness Let $u, v \in L^2(0, T; V) \cap C([0, T]; H)$ a.s. be two adapted solutions. For each $n \geq 1$, we define the stopping time

$$\tau_n = \inf\{t \leq T; |u(t)|^2 \vee |v(t)|^2 \vee \int_0^t (\|u(s)\|^2 + \|v(s)\|^2) \mathrm{d}s \geq n\}.$$

We note that $\tau_n \to \infty$ a.s., as $n \to \infty$. Now we apply Lemma 2.14 to the difference $u(t) - v(t)$, which satisfies

$$u(t) - v(t) = \int_0^t [A(u(s)) - A(v(s))] \mathrm{d}s + \int_0^t [B(u(s)) - B(v(s))] \mathrm{d}W_s.$$

Clearly $M_t = \int_0^t [B(u(s)) - B(v(s))] \mathrm{d}W_s$ is a local martingale, and $\langle M \rangle_t = \int_0^t |B(u(s)) - B(v(s))|_{\mathcal{H}}^2 \mathrm{d}s$. Hence we have

$$|u(t) - v(t)|^2 = 2 \int_0^t \langle A(u(s)) - A(v(s)), u(s) - v(s) \rangle ds$$

$$+ 2 \int_0^t (u(s) - v(s), B(u(s)) - B(v(s))) dW_s$$

$$+ \int_0^t |B(u(s)) - B(v(s))|_{\mathcal{H}}^2 ds.$$

If we write this identity with t replaced by $t \wedge \tau_n = \inf(t, \tau_n)$, it follows from the first part of Remark 2.11 that the stochastic integral

$$\int_0^{t \wedge \tau_n} (u(s) - v(s), B(u(s)) - B(v(s))) dW_s$$

is a martingale with zero mean. Hence taking the expectation and exploiting the monotonicity assumption $(H2)$ yields

$$\mathbb{E}[|u(t \wedge \tau_n) - v(t \wedge \tau_n)|^2] = 2\mathbb{E} \int_0^{t \wedge \tau_n} \langle A(u(s)) - A(v(s)), u(s) - v(s) \rangle ds$$

$$+ \mathbb{E} \int_0^{t \wedge \tau_n} |B(u(s)) - B(v(s))|_{\mathcal{H}}^2 ds$$

$$\leq \lambda \mathbb{E} \int_0^{t \wedge \tau_n} |u(s) - v(s)|^2 ds$$

$$\leq \lambda \mathbb{E} \int_0^t |u(s \wedge \tau_n) - v(s \wedge \tau_n)|^2 ds,$$

hence from Gronwall's Lemma, $u(t \wedge \tau_n) - v(t \wedge \tau_n) = 0$ a.s., for all $0 \leq t \leq T$ and all $n \geq 1$. Uniqueness is proved.

Existence We use a Galerkin approximation. Again, $\{e_k, \ k \geq 1\}$ denotes an orthonormal basis of H, made of elements of V. For each $n \geq 1$, we define

$$V_n = \text{span}\{e_1, e_2, \ldots, e_n\}.$$

The two main steps in the proof of existence are contained in the following two lemmas:

Lemma 2.16. *For all $n \geq 1$, there exists an adapted process $u_n \in C([0, T]; V_n)$ a.s. such that for all $1 \leq k \leq n$,*

$$(u_n(t), e_k) = (u_0, e_k) + \int_0^t \langle A(u_n(s)), e_k \rangle ds + \sum_{\ell=1}^n \int_0^t (B_\ell(u_n(s)), e_k) dW_s^\ell. \quad (2.8)$$

Lemma 2.17.

$$\sup_n \mathbb{E} \left[\sup_{0 \leq t \leq T} |u_n(t)|^2 + \int_0^T \|u_n(t)\|^2 dt \right] < \infty.$$

Let us admit for a moment these two lemmas, and continue the proof of the theorem. Lemma 2.17 tells us that the sequence $\{u_n, \ n \geq 1\}$ is bounded in $L^2(\Omega; C([0,T]; H) \cap L^2(\Omega \times [0,T]; V))$. It then follows from our assumptions that

1. the sequence $\{A(u_n), \ n \geq 1\}$ is bounded in $L^2(\Omega \times [0,T]; V')$;
2. the sequence $\{B(u_n), \ n \geq 1\}$ is bounded in $L^2(\Omega \times [0,T]; \mathcal{H})$.

Hence there exists a subsequence of the original sequence (which, by an abuse of notation, we do not distinguish from the original sequence), such that

$$u_n \rightharpoonup u \text{ in } L^2(\Omega; L^2(0,T;V) \cap L^\infty(0,T;H))$$

$$A(u_n) \rightharpoonup \xi \text{ in } L^2(\Omega \times (0,T);V')$$

$$B(u_n) \rightharpoonup \eta \text{ in } L^2(\Omega \times (0,T);\mathcal{H})$$

weakly (and in fact weakly* in the L^∞ space). It is now easy to let $n \to \infty$ in equation (2.8), and deduce that for all $t \geq 0$, $k \geq 1$,

$$(u(t), e_k) = (u_0, e_k) + \int_0^t \langle \xi(s), e_k \rangle \mathrm{d}s + \sum_{\ell=1}^\infty \int_0^t (\eta_\ell(s), e_k) \mathrm{d}W_s^\ell. \tag{2.9}$$

It thus remains to prove that

Lemma 2.18. *We have the identities* $\xi = A(u)$ *and* $\eta = B(u)$.

We now need to prove the three lemmas.

PROOF OF LEMMA 2.16 If we write the equation for the coefficients of $u_n(t)$ in the basis of V_n, we obtain a usual finite-dimensional Itô equation, to which the classical theory does not quite apply, since the coefficients of that equation need not be Lipschitz. However, several results allow us to treat the present situation, see e.g. Theorem 3.21 in Pardoux and Răşcanu [26]. We shall not discuss this point further, since it is technical, and in all the examples we have in mind, the coefficients of the approximate finite-dimensional equation are locally Lipschitz, which the reader may assume for convenience.

PROOF OF LEMMA 2.17 We first show that

$$\sup_n \left[\sup_{0 \leq t \leq T} \mathbb{E}(|u_n(t)|^2) + \mathbb{E}\int_0^T \|u_n(s)\|^2 \mathrm{d}s \right] < \infty. \tag{2.10}$$

From the equation (2.8) and Itô's formula, we deduce that for all $1 \leq k \leq n$,

$$(u_n(t), e_k)^2 = (u_0, e_k)^2 + 2\int_0^t (u_n(s), e_k)\langle A(u_n(s)), e_k \rangle \mathrm{d}s$$

$$+ 2\sum_{\ell=1}^n \int_0^t (u_n(s), e_k)(B_\ell(u_n(s)), e_k)\mathrm{d}W_s^\ell + \sum_{\ell=1}^n \int_0^t (B_\ell(u_n(s)), e_k)^2 \mathrm{d}s.$$

Summing from $k = 1$ to $k = n$, we obtain

$$|u_n(t)|^2 = \sum_{k=1}^{n} (u_0, e_k)^2 + 2 \int_0^t \langle A(u_n(s)), u_n(s) \rangle \mathrm{d}s$$

$$+ 2 \sum_{\ell=1}^{n} \int_0^t (B_\ell(u_n(s)), u_n(s)) \mathrm{d}W_s^\ell + \sum_{\ell=1}^{n} \sum_{k=1}^{n} \int_0^t (B_\ell(u_n(s)), e_k)^2 \mathrm{d}s,$$

$$(2.11)$$

from which we deduce that

$$|u_n(t)|^2 \leq |u_0|^2 + 2 \int_0^t \langle A(u_n(s)), u_n(s) \rangle \mathrm{d}s$$

$$(2.12)$$

$$+ 2 \sum_{\ell=1}^{n} \int_0^t (B_\ell(u_n(s)), u_n(s)) \mathrm{d}W_s^\ell + \int_0^t |B(u_n(s))|_{\mathcal{H}}^2 \mathrm{d}s.$$

Now we take the expectation in the above inequality:

$$\mathbb{E}(|u_n(t)|^2) \leq |u_0|^2 + 2\mathbb{E} \int_0^t \langle A(u_n(s)), u_n(s) \rangle \mathrm{d}s + \mathbb{E} \int_0^t |B(u_n(s))|_{\mathcal{H}}^2 \mathrm{d}s,$$

and combine the resulting inequality with the assumption $(H1)$, yielding

$$\mathbb{E}\left(|u_n(t)|^2 + \alpha \int_0^t \|u_n(s)\|^2 \mathrm{d}s\right) \leq |u_0|^2 + \lambda \mathbb{E} \int_0^t |u_n(s)|^2 \mathrm{d}s + \nu t. \qquad (2.13)$$

Combining with Gronwall's Lemma, we conclude that

$$\sup_{n} \sup_{0 \leq t \leq T} \mathbb{E}(|u_n(t)|^2) < \infty,$$

and combining the last two inequalities, we deduce that

$$\sup_{n} \mathbb{E} \int_0^T \|u_n(t)\|^2 \mathrm{d}t < \infty. \qquad (2.14)$$

The estimate (2.10) follows from (2.13) and (2.14). We now take the sup over t in (2.12), yielding

$$\sup_{0 \leq t \leq T} |u_n(t)|^2 \leq |u_0|^2 + 2 \int_0^T |\langle A(u_n(s)), u_n(s) \rangle| \mathrm{d}s$$

$$+ 2 \sup_{0 \leq t \leq T} \left| \sum_{\ell=1}^{n} \int_0^t (B_\ell(u_n(s)), u_n(s)) \mathrm{d}W_s^\ell \right| + \int_0^T |B(u_n(s))|_{\mathcal{H}}^2 \mathrm{d}s.$$

$$(2.15)$$

Now the Burkholder–Davis–Gundy inequality tells us that

$$\mathbb{E}\left[2\sup_{0\leq t\leq T}\left|\sum_{\ell=1}^{n}\int_{0}^{t}(B_{\ell}(u_{n}(s)),u_{n}(s))\mathrm{d}W_{s}^{\ell}\right|\right]$$

$$\leq c\mathbb{E}\sqrt{\sum_{\ell=1}^{n}\int_{0}^{T}(B_{\ell}(u_{n}(t)),u_{n}(t))^{2}\mathrm{d}t}$$

$$\leq c\mathbb{E}\left[\sup_{0\leq t\leq T}|u_{n}(t)|\sqrt{\int_{0}^{T}|B(u_{n}(t))|_{\mathcal{H}}^{2}\mathrm{d}t}\right]$$

$$\leq \frac{1}{2}\mathbb{E}\left(\sup_{0\leq t\leq T}|u_{n}(t)|^{2}\right)+\frac{c^{2}}{2}\mathbb{E}\int_{0}^{T}|B(u_{n}(t))|_{\mathcal{H}}^{2}\mathrm{d}t.$$

Combining (2.15) with the assumption ($H1$) and this last inequality, we deduce that

$$\mathbb{E}\left(\sup_{0\leq t\leq T}|u_{n}(t)|^{2}\right)\leq 2|u_{0}|^{2}+c'\mathbb{E}\int_{0}^{T}(1+|u_{n}(t)|^{2}\mathrm{d}t.$$

The result follows from this and (2.10).

PROOF OF LEMMA 2.18 We are going to exploit the monotonicity assumption ($H2$), which for simplicity we assume to hold with $\lambda=0$ (this is in fact not necessary, but is also not a restriction). ($H2$) with $\lambda=0$ implies that for all $v\in L^{2}(\Omega\times(0,T);V)$ and all $n\geq 1$,

$$2\mathbb{E}\int_{0}^{T}\langle A(u_{n}(t)-A(v(t)),u_{n}(t)-v(t)\rangle\mathrm{d}t+\mathbb{E}\int_{0}^{T}|B(u_{n}(t))-B(v(t))|_{\mathcal{H}}^{2}\mathrm{d}t\leq 0.$$
(2.16)

Weak convergence implies that

$$\int_{0}^{T}\langle A(u_{n}(t)),v(t)\rangle\mathrm{d}t\rightharpoonup\int_{0}^{T}\langle\xi(t),v(t)\rangle\mathrm{d}t,$$

$$\int_{0}^{T}\langle A(v(t)),u_{n}(t)\rangle\mathrm{d}t\rightharpoonup\int_{0}^{T}\langle A(v(t)),u(t)\rangle\mathrm{d}t,$$
(2.17)

$$\int_{0}^{T}(B(u_{n}(t)),B(v(t)))_{\mathcal{H}}\mathrm{d}t\rightharpoonup\int_{0}^{T}(\eta(t),B(v(t)))_{\mathcal{H}}\mathrm{d}t$$

in $L^{2}(\Omega)$ weakly. Suppose we have in addition the inequality

$$2\mathbb{E}\int_{0}^{T}\langle\xi(t),u(t)\rangle\mathrm{d}t+\mathbb{E}\int_{0}^{T}|\eta(t)|_{\mathcal{H}}^{2}\mathrm{d}t$$

$$\leq \liminf_{n\to\infty}\mathbb{E}\left[2\int_{0}^{T}\langle A(u_{n}(t)),u_{n}(t)\rangle\mathrm{d}t+\int_{0}^{T}|B(u_{n}(t)|_{\mathcal{H}}^{2}\mathrm{d}t\right].$$
(2.18)

It follows from (2.16), (2.17) and (2.18) that for all $v\in L^{2}(\Omega\times(0,T);V)$,

$$2\mathbb{E}\int_{0}^{T}\langle\xi(t)-A(v(t)),u(t)-v(t)\rangle\mathrm{d}t+\mathbb{E}\int_{0}^{T}|\eta(t)-B(v(t))|_{\mathcal{H}}^{2}\mathrm{d}t\leq 0. \quad (2.19)$$

We first choose $v = u$ in (2.19), and deduce that $\eta \equiv B(u)$. Moreover, (2.19) implies that

$$\mathbb{E} \int_0^T \langle \xi(t) - A(v(t)), u(t) - v(t) \rangle dt \le 0.$$

Next we choose $v(t) = u(t) - \theta w(t)$, with $\theta > 0$ and $w \in L^2(\Omega \times (0,T); V)$. After division by θ, we obtain the inequality

$$\mathbb{E} \int_0^T \langle \xi(t) - A(u(t) - \theta w(t)), w(t) \rangle dt \le 0.$$

We now let $\theta \to 0$, and thanks to the assumption $(H4)$, we deduce that

$$\mathbb{E} \int_0^T \langle \xi(t) - A(u(t)), w(t) \rangle dt \le 0, \quad \forall w \in L^2(\Omega \times (0,T); V).$$

It clearly follows that $\xi \equiv A(u)$.

It remains to establish the inequality (2.18). It follows from (2.11) that

$$2\mathbb{E} \int_0^T \langle A(u_n(t)), u_n(t) \rangle dt + \mathbb{E} \int_0^T |B(u_n(t)|_{\mathcal{H}}^2 dt \ge \mathbb{E} \left[|u_n(T)|^2 - |u_n(0)|^2 \right],$$

and from Lemma 2.14 applied to $u(t)$ satisfying (2.9) that

$$2\mathbb{E} \int_0^T \langle \xi(t), u(t) \rangle dt + \mathbb{E} \int_0^T |\eta(t)|_{\mathcal{H}}^2 dt = \mathbb{E} \left[|u(T)|^2 - |u_0|^2 \right].$$

Hence (2.18) is a consequence of the inequality

$$\mathbb{E} \left[|u(T)|^2 - |u_0|^2 \right] \le \liminf_{n \to \infty} \mathbb{E} \left[|u_n(T)|^2 - |u_n(0)|^2 \right].$$

But clearly $u_n(0) = \sum_{k=1}^n (u_0, e_k) e_k \to u_0$ in H. Hence the result will follow from the convexity of the mapping $\rho \to \mathbb{E}(|\rho|^2)$ from $L^2(\Omega, \mathcal{F}_T, \mathbb{P}; H)$ into $\mathbb{R}$, provided we show that $u_n(T) \to u(T)$ in $L^2(\Omega, \mathcal{F}_T, \mathbb{P}, H)$ weakly. Since the sequence $\{u_n(T), n \ge 1\}$ is bounded in $L^2(\Omega, \mathcal{F}_T, \mathbb{P}, H)$, we can w.l.o.g. assume that the subsequence has been chosen in such a way that $u_n(T)$ converges weakly in $L^2(\Omega, \mathcal{F}_T, \mathbb{P}, H)$ as $n \to \infty$. On the other hand, for any n_0 and $v \in V_{n_0}$, whenever $n \ge n_0$,

$$(u_n(T), v) = (u_0, v) + \int_0^T \langle A(u_n(t)), v \rangle dt + \sum_{\ell=1}^n \int_0^T (B_\ell(u_n(t)), v) dW_t^\ell.$$

The right-hand side converges weakly in $L^2(\Omega, \mathcal{F}_T, \mathbb{P}; \mathbb{R})$ towards

$$(u_0, v) + \int_0^T \langle \xi(t), v \rangle dt + \sum_{\ell=1}^\infty \int_0^T (\eta_\ell(t), v) dW_t^\ell = (u(T), v).$$

The result follows.

2.4.2 Examples

A simple example

We start with a simple example which will illustrate the coercivity condition. Consider the following parabolic "bilinear" SPDE with space dimension equal to one, driven by a one-dimensional Wiener process, namely

$$\frac{\partial u}{\partial t}(t,x) = \frac{1}{2}\frac{\partial^2 u}{\partial x^2}(t,x) + \theta \frac{\partial u}{\partial x}(t,x)\frac{dW}{dt}(t); \, |u(0,x) = u_0(x).$$

The coercivity condition, when applied to this SPDE, yields the restriction $|\theta| < 1$. Under this assumption, the solution, starting from $u_0 \in H$, is in V for a.e. $t > 0$, i.e. we have the regularization effect of a parabolic equation.

When $\theta = 1$ (resp. $\theta = -1$), we deduce from Itô's formula the explicit solution $u(t,x) = u_0(x + W(t))$ (resp. $u(t,x) = u_0(x - W(t))$). It is easily seen that in this case the regularity in x of the solution is the same at each time $t > 0$ as it is at time 0. This should not be considered as a parabolic equation, but rather as a first-order hyperbolic equation.

What happens if $|\theta| > 1$? We suspect that solving the SPDE in that case raises the same type of difficulty as solving a parabolic equation (like the heat equation) backward in time.

Note that the above equation is equivalent to the following SPDE in the Stratonovich sense

$$\frac{\partial u}{\partial t}(t,x) = \frac{1-\theta^2}{2}\frac{\partial^2 u}{\partial x^2}(t,x) + \theta \frac{\partial u}{\partial x}(t,x) \circ \frac{dW}{dt}(t); \, u(0,x) = u_0(x).$$

Zakai's equation

We look at the equation for the density p in the above Subsection 1.2.5. We assume that the following are bounded functions defined on $\mathbb{R}^d$: $a, b, h, g, \frac{\partial a_{ij}}{\partial x_j}, \frac{\partial g_{i\ell}}{\partial x_i}$, for all $1 \le i, j \le d, 1 \le \ell \le k$. The equation for p is of the form

$$\frac{\partial p}{\partial t}(t,x) = Ap(t,x) + \sum_{\ell=1}^{k} B_\ell p(t,x)\frac{dW_\ell}{dt}(t), \quad \cdot$$

where

$$Au = \frac{1}{2}\sum_{i,j}\frac{\partial}{\partial x_i}\left(a_{ij}\frac{\partial u}{\partial x_j}\right) + \sum_i \frac{\partial}{\partial x_i}\left(\left(\sum_j \frac{1}{2}\frac{\partial a_{ij}}{\partial x_j} - b_i\right)u\right)$$

and

$$B_\ell = -\sum_i g_{i\ell}\frac{\partial u}{\partial x_i} + \left(h_\ell - \sum_i \frac{\partial g_{i\ell}}{\partial x_i}\right)u.$$

We note that

$$
2\langle Au, u\rangle + \sum_{\ell=1}^{k} |B_\ell u|^2 = \sum_{i,j} \int_{\mathbb{R}^d} (gg^* - a)_{ij}(x)\frac{\partial u}{\partial x_i}(x)\frac{\partial u}{\partial x_j}(x)\mathrm{d}x
$$

$$
+ \sum_i \int_{\mathbb{R}^d} c_i(x)\frac{\partial u}{\partial x_i}(x)u(x)\mathrm{d}x + \int_{\mathbb{R}^d} d(x)u^2(x)\mathrm{d}x.
$$

Whenever $ff^*(x) > \beta I > 0$ for all $x \in \mathbb{R}^d$, the coercivity assumption is satisfied for any $\alpha < \beta$, some $\lambda > 0$ and $\nu = 0$. Note that it is very natural that the ellipticity assumption concerns the matrix ff^*. Indeed, in the particular case where $h \equiv 0$, we observe the Wiener process W, so the uncertainty in the conditional law of X_t given $\mathcal{F}_t^Y$ depends on the diffusion matrix ff^* only. The case without the restriction that ff^* be elliptic can be studied, but we need some more regularity of the coefficients.

Nonlinear examples

One can always add a term of the form

$$
f_1(t,x,u) + f_2(t,x,u)
$$

to $A(u)$, provided $u \to f_1(t,x,u)$ is decreasing for all (t,x), and $f_2(t,x,u)$ is Lipschitz in u, with a uniform Lipschitz constant independent of (t,x). Note that a typical decreasing f_1 is given by

$$
f_1(t,x,u) = -c(t,x)|u|^{p-2}u, \quad \text{provided that } c(t,x) \geq 0.
$$

Similarly, one can add to $B(u)$ a term $g(t,x,u)$, where g has the same property as f_2.

Another nonlinear example

The following operator (with $p > 2$)

$$
A(u) = \sum_{i=1}^{d} \frac{\partial}{\partial x_i}\left(\left|\frac{\partial u}{\partial x_i}\right|^{p-2}\frac{\partial u}{\partial x_i}\right) - |u|^{p-2}u
$$

possesses all the required properties, if we let $H = L^2(\mathbb{R}^d)$,

$$
V = W^{1,p}(\mathbb{R}^d) = \{u \in L^p(\mathbb{R}^d), \frac{\partial u}{\partial x_i} \in L^p(\mathbb{R}^d), i = 1,\ldots,d\}
$$

and $V' = W^{-1,q}(\mathbb{R}^d)$, where $1/p + 1/q = 1$.

2.4.3 Coercive SPDEs with compactness

We keep the assumptions $(H1)$ and $(H3)$ from the previous subsection, and we add the following conditions.

Sublinear growth of B

$$(H5) \begin{cases} \exists c, \delta > 0 \text{ such that } \forall u \in V, \\ |B(u)|_{\mathcal{H}} \leq c(1 + \|u\|^{1-\delta}) \end{cases}$$

Compactness

$$(H6) \quad \text{The injection from } V \text{ into } H \text{ is compact.}$$

Continuity

$$(H7) \begin{cases} u \to A(u) \text{ is continuous from } V_{\text{weak}} \cap H \text{ into } V'_{\text{weak}} \\ u \to B(u) \text{ is continuous from } V_{\text{weak}} \cap H \text{ into } \mathcal{H} \end{cases}$$

We now want to formulate our SPDE as a martingale problem. We choose

$$\Omega = C([0,T]; H_{\text{weak}}) \cap L^2(0,T;V) \cap L^2(0,T;H),$$

which we equip with the sup of the topology of uniform convergence with values in H equipped with its weak topology, the weak topology of $L^2(0,T;V)$, and the strong topology of $L^2(0,T;H)$. Moreover, we let $\mathcal{F}$ be the associated Borel σ-field. For $0 \leq t \leq T$, let Ω_t denote the same space as Ω, but with T replaced by t, and Π_t be the projection from Ω into Ω_t, which to a function defined on the interval $[0,T]$ associates its restriction to the interval $[0,t]$. Now $\mathcal{F}_t$ will denote the smallest sub-σ-field of $\mathcal{F}$, which makes the projection Π_t measurable, when Ω_t is equipped with its own Borel σ-field. From now on, in this subsection, we define $u(t,\omega) = \omega(t)$.

Definition 2.19. A probability $\mathbb{P}$ on $(\Omega, \mathcal{F})$ is a solution to the martingale problem associated with the SPDE (2.6) whenever

(i) $\mathbb{P}(u(0) = u_0) = 1$;
(ii) the process

$$M_t := u(t) - u(0) - \int_0^t A(u(s))\mathrm{d}s$$

is a continuous H-valued $\mathbb{P}$-martingale with associated increasing process

$$\langle\langle M \rangle\rangle_t = \int_0^t B(u(s))B^*(u(s))\mathrm{d}s.$$

There are several equivalent formulations of (ii). Let us give the formulation which we will actually use below. Let $\{e_i,\ i = 1, 2, \ldots\}$ be an orthonormal basis of H, with $e_i \in V, \forall i \geq 1$.

(ii)' For all $i \geq 1$, $\varphi \in C_b^2(\mathbb{R})$, and continuous, bounded and $\mathcal{F}_s$-measurable mappings Φ_s ($0 \leq s \leq t$) from Ω into $\mathbb{R}$,

$$\mathbb{E}_{\mathbb{P}}\left((M_t^{i,\varphi} - M_s^{i,\varphi})\Phi_s\right) = 0, \quad \text{where}$$

$$M_t^{i,\varphi} = \varphi[(u(t), e_i)] - \varphi[(u_0, e_i)] - \int_0^t \varphi'[(u(s), e_i)]\langle A(u(s)), e_i\rangle \mathrm{d}s$$
$$+ \frac{1}{2}\int_0^t \varphi''[(u(s), e_i)](BB^*(u(s))e_i, e_i)\mathrm{d}s.$$

This formulation of a martingale problem for solving stochastic differential equations was first introduced by Stroock and Varadhan for solving finite-dimensional SDEs, and by Viot in his thesis (1976) for solving SPDEs. It is his results which we present here.

We first note that if we have a solution to the SPDE, its probability law on Ω solves the martingale problem. Conversely, if we have a solution to the martingale problem, then we have a probability space $(\Omega, \mathcal{F}, \mathbb{P})$, and an H-valued process $\{u(t), 0 \leq t \leq T\}$ defined on it, with trajectories in $L^2(0, T; V)$, such that

$$u(t) = u_0 + \int_0^t A(u(s))\mathrm{d}s + M_t,$$

where $\{M_t, 0 \leq t \leq T\}$ is a continuous H-valued martingale, and

$$\langle\langle M \rangle\rangle_t = \int_0^t B(u(s))B^*(u(s))\mathrm{d}s.$$

It follows from a representation theorem similar to a well-known result in finite dimensions that there exists, possibly on a larger probability space, a Wiener process $\{W(t), t \geq 0\}$ such that (2.6) holds. A solution of the martingale problem is called a *weak solution* of the SPDE, in the sense that one can construct a pair $\{(u(t), W(t)), t \geq 0\}$ such that the second element is a Wiener process, and the first solves the SPDE driven by the second, while until now we have given ourselves $\{W(t), t \geq 0\}$, and we have found the corresponding solution $\{u(t), t \geq 0\}$.

We next note that whenever a SPDE is such that it admits at most one strong solution (i.e., to each given Wiener process W, we can associate at most one solution u of the SPDE driven by W), then the martingale problem also has at most one solution.

We now prove the following theorem.

Theorem 2.20. *Under the assumptions (H1), (H3), (H5), (H6) and (H7), there exists a solution $\mathbb{P}$ to the martingale problem, i.e. which satisfies (i) and (ii).*

PROOF: We start with the same Galerkin approximation which we have used before. Again $\{e_1, \ldots, e_n, \ldots\}$ is an orthonormal basis of H, with each $e_n \in V$,

$$V_n = \mathrm{span}\{e_1, \ldots, e_n\}$$

$$\pi_n = \text{the orthogonal projection operator in } H \text{ upon } V.$$

We first note that for each $n \geq 1$, there exists a probability measure $\mathbb{P}_n$ on $(\Omega, \mathcal{F})$ such that

$(0)_n$ $\mathrm{Supp}(\mathbb{P}_n) \subset C([0,T]; V_n)$;
$(i)_n$ $\mathbb{P}_n(u(0) = \pi_n u_0) = 1$
$(ii)_n$ $\forall i \leq n, \varphi \in C_b^2(\mathbb{R}), 0 \leq s \leq t \leq T,$

$$\mathbb{E}_n\left((M_t^{i,\varphi} - M_s^{i,\varphi})\Phi_s\right) = 0, \quad \text{where}$$

$\{M_t^{i,\varphi}\}$ and Φ_s are defined exactly as in condition (ii) and (ii)' of Definition 2.19.

Indeed, the existence of each P_n is obtained by solving finite-dimensional martingale problems (or finite-dimensional SDEs). This works without any serious difficulty, and we take this result for granted.

Let us accept for a moment the following lemma.

Lemma 2.21. *The sequence of probability measures* $\{\mathbb{P}_n, \; n = 1, 2, \ldots\}$ *on* Ω *is tight.*

We shall admit the fact (which has been proved by M. Viot in his thesis) that Prohorov's theorem is valid in the space Ω. This is not obvious, since Ω is not a Polish space, but it is true. Hence we can extract from the sequence $\{P_n, \; n = 1, 2, \ldots\}$ a subsequence, which as an abuse of notation we still denote by $\{P_n\}$, such that $\mathbb{P}_n \Rightarrow \mathbb{P}$. Now $\mathbb{P}$ clearly satisfies (i), and for each $0 < s < t$, the mapping

$$\omega \to (M_t^{i,\varphi}(\omega) - M_s^{i,\varphi}(\omega))\Phi_s(\omega)$$

is continuous from Ω into $\mathbb{R}$. Moreover, it follows from the coercivity assumption $(H1)$ that the estimate

$$\sup_n \mathbb{E}_n\left[\sup_{0 \leq t \leq T} |u(t)|^2 + \int_0^T \|u(t)\|^2 dt\right] < \infty \tag{2.20}$$

from Lemma 2.17 is still valid. Now this plus the conditions $(H3)$ and $(H5)$ implies that there exists some $p > 1$ (the exact value of p depends upon the value of δ in condition $(H5)$) such that

$$\sup_n \mathbb{E}_n\left[|M_t^{i,\varphi} - M_s^{i,\varphi}|^p\right] < \infty.$$

Hence

$$\mathbb{E}_n\left((M_t^{i,\varphi} - M_s^{i,\varphi})\Phi_s\right) \to \mathbb{E}\left((M_t^{i,\varphi} - M_s^{i,\varphi})\Phi_s\right),$$

and condition (ii) is met. It remains to proceed to the proof of Lemma 2.21.

PROOF OF LEMMA 2.21 (SKETCH): Let us denote by

- τ_1 the weak topology on $L^2(0, T; V)$,
- τ_2 the uniform topology on $C([0, T]; H_{weak})$,
- τ_3 the weak topology of $L^2(0, T; H)$.

It suffices to show that the sequence $\{\mathbb{P}_n, n \geq 1\}$ is τ_i-tight successively for $i = 1, 2, 3$.

1. τ_1-tightness. We choose

$$K_1 = \{u, \int_0^T \|u(t)\|^2 dt \leq k\}.$$

K_1 is relatively compact for the weak topology τ_1, since it is a bounded subset of $L^2(0, T; V)$, which is a reflexive Banach space. But it follows from (2.20) that there exists a $c \in \mathbb{R}$ such that

$$\mathbb{E}_n \int_0^T \|u(t)\|^2 dt \leq c,$$

hence from Chebychev's inequality

$$\mathbb{P}_n \left(\int_0^T \|u(t)\|^2 dt > k \right) \leq \frac{c}{k},$$

K_1 possesses the required properties, provided we choose k large enough.

2. τ_2-tightness. We want to find K_2 in such a way that for $h \in H$ with $|h| = 1$, the set of functions

$$\{t \rightarrow (u(t), h), \quad u \in K_2\}$$

is a compact subset of $C([0, T])$. From (2.20), there exists a $c \in \mathbb{R}$ such that

$$\mathbb{E}_n \left(\sup_{0 \leq t \leq T} |u(t)|^2 \right) \leq c.$$

So it is sufficient to get that for any $v \in V$ with $\|v\| = 1$, the set of functions

$$\{t \rightarrow (u(t), v), \quad u \in K_2\}$$

is a compact subset of $C([0, T])$. Now $\sup_{0 \leq t \leq T} |(u(t), v)|$ is well controlled. So, using the Arzela–Ascoli theorem, it suffices to control the modulus of continuity of $\{t \rightarrow (u(t), v)\}$ uniformly in $u \in K_2$. But

$$(u(t), v) = (u_0, v) + \int_0^t \langle A(u(s)), v \rangle ds + M_t^v, \text{ and}$$

$$\mathbb{E}_n \left| \int_s^t \langle A(u(r)), v \rangle dr \right| \leq \|e_i\| \sqrt{t-s} \sqrt{\mathbb{E}_n \int_0^T \|A(u(r))\|_*^2 dr}$$

$$\leq c \|v\| \sqrt{t-s},$$

$$\mathbb{E}_n\left(\sup_{s\leq r\leq t}|M_r^\nu - M_s^\nu|^{2p}\right) \leq c_p|\nu|^p\mathbb{E}_n\left(\left|\int_s^t(BB^*(u(r))e_i,e_i)\mathrm{d}r\right|^p\right)$$

$$\leq c_p|\nu|^p(t-s)^{p\delta}\left(\mathbb{E}_n\int_0^T(1+\|u(r)\|^2)\mathrm{d}r\right)^{p(1-\delta)},$$

for all $p > 0$, δ being the constant from condition $(H5)$.

3. τ_3-tightness. We just saw in fact that we can control the modulus of continuity of $\{t \to u(t)\}$ as a V'-valued function under $\mathbb{P}_n$. Recall the bound

$$\mathbb{E}_n\int_0^T\|u(t)\|^2\mathrm{d}t \leq c.$$

It remains to exploit the next lemma.

Lemma 2.22. *Given that the injection from V into H is compact, from any sequence $\{u_n, n \geq 1\}$ which is both bounded in $L^2(0,T;V)\cap L^\infty(0,T;H)$ and equicontinuous as V'-valued functions, and such that the sequence $\{u_n(0)\}$ converges strongly in H, one can extract a subsequence which converges in $L^2(0,T;H)$ strongly.*

We first prove the following

Lemma 2.23. *To each $\varepsilon > 0$, we can associate $c(\varepsilon) \in \mathbb{R}$ such that for all $v \in V$,*

$$|v| \leq \varepsilon\|v\| + c(\varepsilon)\|v\|_*.$$

PROOF: If the result was not true, then one could find $\varepsilon > 0$ and a sequence $\{v_n, n \geq 1\} \subset V$ such that for all $n \geq 1$,

$$|v_n| \geq \varepsilon\|v_n\| + n\|v_n\|_*.$$

We define $u_n = |v_n|^{-1}v_n$. Then we have that

$$1 = |u_n| \geq \varepsilon\|u_n\| + n\|u_n\|_*.$$

This last inequality shows both that the sequence $\{u_n,\ n \geq 1\}$ is bounded in V, and converges to 0 in V'. Hence, from the compactness of the injection from V into H, $u_n \to u$ in H strongly, and necessarily $u = 0$. But this contradicts the fact that $|u_n| = 1$ for all n. $\qquad\square$

PROOF OF LEMMA 2.22: From the equicontinuity in V' and the fact that $u_n(0) \to u_0$ in H, there is a subsequence which converges in $C([0,T];V')$, hence also in $L^2(0,T;V')$, to u, and clearly $u \in L^2(0,T;V)$. Now from Lemma 2.23, to each $\varepsilon > 0$, we can associate $c'(\varepsilon)$ such that

$$\int_0^T|u_n(t)-u(t)|^2\mathrm{d}t \leq \varepsilon\int_0^T\|u_n(t)-u(t)\|^2\mathrm{d}t + c'(\varepsilon)\int_0^T\|u_n(t)-u(t)\|_*^2\mathrm{d}t$$

$$\leq \varepsilon C + c'(\varepsilon)\int_0^T\|u_n(t)-u(t)\|_*^2\mathrm{d}t,$$

$$\limsup_n \int_0^T |u_n(t) - u(t)|^2 dt \le C\varepsilon,$$

and the result follows from the fact that ε can be chosen arbitrarily small. $\qquad\Box$

2.5 Semilinear SPDEs

We want now to concentrate on the following class of SPDEs

$$\begin{cases} \dfrac{\partial u}{\partial t}(t,x) = \dfrac{1}{2} \sum_{ij} \dfrac{\partial}{\partial x_j}\left(a_{ij}(t,x)\dfrac{\partial u}{\partial x_i}\right)(t,x) + \sum_i b_i(t,x)\dfrac{\partial u}{\partial x_i}(t,x) \\[3mm] \qquad + f(t,x;u(t,x)) \\[3mm] \qquad + \displaystyle\sum_k \left(\sum_i g_{ki}(t,x)\dfrac{\partial u}{\partial x_i}(t,x) + h_k(t,x;u(t,x))\right)\dfrac{dW^k}{dt}(t) \\[3mm] u(0,x) = u_0(x) \end{cases} \quad (2.21)$$

Under the following standard assumptions

- $\exists \alpha > 0$ such that $\bar{a} = a - \sum_k g_k.g_k. \ge \alpha I$;

- $2[f(t,x;r) - f(t,x;r')](r-r') + \sum_k |h_k(t,x;r) - h_k(t,x;r')|^2 \le \lambda |r-r'|^2$;

- $r \longrightarrow f(t,x;r)$ is continuous;

- $rf(t,x;r) + \sum_k |h_k(t,x;r)|^2 \le C(1+|r|^2)$,

equation (2.21) has a unique solution with trajectories in $C([0,T];L^2(\mathbb{R}^d)) \cap L^2(0,T;H^1(\mathbb{R}^d))$, as follows from Theorem 2.13.

Let us now give conditions under which the solution remains nonnegative.

Theorem 2.24. *Assume that $u_0(x) \ge 0$, for a.e. x, and for a.e. t and x, $f(t,x;0) \ge 0$, $h_k(t,x;0) = 0$, for all k. Then*

$$u(t,x) \ge 0, \quad \forall t \ge 0, \ x \in \mathbb{R}^d.$$

PROOF: Let us consider the new equation (below $u^+(t,x) := \sup(u(t,x),0)$)

$$\begin{cases} \dfrac{\partial u}{\partial t}(t,x) = \dfrac{1}{2} \sum_{ij} \dfrac{\partial}{\partial x_j}\left(a_{ij}(t,x)\dfrac{\partial u}{\partial x_i}\right)(t,x) \\[3mm] \qquad + \displaystyle\sum_i b_i(t,x)\dfrac{\partial u^+}{\partial x_i}(t,x) + f(t,x;u^+(t,x)) \\[3mm] = \displaystyle\sum_k \left(\sum_i g_{ki}(t,x)\dfrac{\partial u}{\partial x_i}(t,x) + h_k(t,x;u^+(t,x))\right)\dfrac{dW^k}{dt}(t) \end{cases} \quad (2.22)$$

Existence and uniqueness for this new equation follows almost the same arguments as for equation (2.21). We exploit the fact that the mapping $r \to r^+$ is Lipschitz. Moreover, we can w.l.o.g. assume that the $\partial b_i / \partial x_i$'s are bounded functions, since the general result will follow from the theorem by taking the limit along a converging sequence of smooth coefficients. However, it is not hard to show that, with this additional assumption, the mapping

$$u \to \sum_i b_i(t,x) \frac{\partial u^+}{\partial x_i}$$

is compatible with the coercivity and monotonicity of the pair of operators appearing in (2.22). If we can show that the solution of (2.22) is nonnegative, then it is the unique solution of (2.21), which then is nonnegative.

Let $\varphi \in C^2(\mathbb{R})$ be convex and such that

$$\begin{cases} \bullet \ \varphi(r) = 0, & \text{for } r \geq 0; \\ \bullet \ \varphi(r) > 0, & \text{for } r < 0; \\ \bullet \ 0 \leq \varphi(r) \leq Cr^2 \quad \forall r; \\ \bullet \ -c|r| \leq \varphi'(r) \leq 0 \quad \forall r; \\ \bullet \ 0 \leq \varphi''(r) \leq C \quad \forall r. \end{cases}$$

Intuitively, φ is a regularization of $(r^-)^2$. Let now $\Phi : L^2(\mathbb{R}) \to \mathbb{R}$ be defined by

$$\Phi(u) = \int_{\mathbb{R}^d} \varphi(u(x)) dx.$$

We have $\Phi'(h) = \varphi'(h(\cdot))$, which is well defined as an element of $L^2(\mathbb{R}^d)$, since $|\varphi'(x)| \leq c|x|$, and $\Phi''(h) = \varphi''(h(\cdot))$, and it belongs to $\mathcal{L}(L^2(\mathbb{R}^d))$, since $|\varphi''(x)| \leq C$. We let

$$Au = \frac{1}{2} \sum_{ij} \frac{\partial}{\partial x_j} \left(a_{ij} \frac{\partial u}{\partial x_i} \right) + \sum_i b_i(t,x) \frac{\partial u^+}{\partial x_i} + f(u^+)$$

$$B_k u = \sum_i g_{ki} \frac{\partial u}{\partial x_i} + h_k(u^+).$$

It follows from the Itô formula from Lemma 2.15 that

$$\Phi(u(t)) = \Phi(u_0) + \int_0^t \langle A(u(s)), \varphi'(u(s)) \rangle ds$$

$$+ \sum_k \int_0^t (B_k(u(s)), \varphi'(u(s))) \, dW_s^k$$

$$+ \frac{1}{2} \sum_k \int_0^t (B_k(u(s)), \varphi''(u(s)) B_k(u(s))) \, ds,$$

Now $\Phi(u_0) = 0$, and

$$\mathbb{E}\Phi(u(t)) = -\frac{1}{2}\mathbb{E}\int_0^t ds \int_{\mathbb{R}^d} dx \left(\varphi''(u)\langle \overline{a}\nabla u, \nabla u\rangle\right)(s,x)$$

$$+ \mathbb{E}\int_0^t ds \int_{\mathbb{R}^d} dx\varphi'(u)[f(u^+) + \sum_i b_i \frac{\partial u^+}{\partial x_i}](s,x)$$

$$+ \sum_k \mathbb{E}\int_0^t ds \int_{\mathbb{R}^d} dx\varphi''(u)h_k(u^+)[\frac{1}{2}h_k(u^+) + g_{kj}\frac{\partial u}{\partial x_j}](s,x)$$

$$\leq 0,$$

where we have used the following lemma.

Lemma 2.25. *Whenever* $u \in H^1(\mathbb{R}^d)$, $u^+ \in H^1(\mathbb{R}^d)$, *and moreover*

$$\frac{\partial u^+}{\partial x_i}(x)\mathbf{1}_{\{u<0\}}(x) = 0, \ dx \ a.e. \ , \forall 1 \leq i \leq d.$$

If we admit this lemma for a moment, we note that we have proved that for any $t \geq 0$, $\mathbb{E}\Phi(u(t)) = 0$, i.e. $\Phi(u(t)) = 0$ a.s., and in fact $u(t,x) \geq 0$, dx a.e., a.s., $\forall t$.

PROOF OF LEMMA 2.25: We define a sequence of approximations of the function $r \to r^+$ of class C^1:

$$\varphi_n(r) = \begin{cases} 0, & \text{if } r < 0; \\ nr^2/2, & \text{if } 0 < r < 1/n; \\ r - 1/2n, & \text{if } r > 1/n. \end{cases}$$

Clearly, $\varphi_n(r) \to r^+$, and $\varphi'_n(r) \to \mathbf{1}_{\{r>0\}}$, as $n \to \infty$. For $u \in H^1(\mathbb{R}^d)$, let $u_n(x) = \varphi_n(u(x))$. Then $u_n \in H^1(\mathbb{R}^d)$, and

$$\frac{\partial u_n}{\partial x_i} = \varphi'_n(u)\frac{\partial u}{\partial x_i}.$$

It is easily seen that the two following convergences hold in $L^2(\mathbb{R}^d)$:

$$u_n \to u^+, \qquad \frac{\partial u_n}{\partial x_i} \to \mathbf{1}_{\{u>0\}}\frac{\partial u}{\partial x_i}.$$

This proves the lemma.   □

By a similar argument, one can also prove a comparison theorem.
Let v be the solution of a slightly different SPDE

$$\begin{cases} \frac{\partial v}{\partial t}(t,x) = \frac{1}{2}\sum_{ij}\frac{\partial}{\partial x_j}\left(a_{ij}(t,x)\frac{\partial v}{\partial x_i}\right)(t,x) + \sum_i b_i(t,x)\frac{\partial v}{\partial x_i}(t,x) \\ \qquad + F(t,x;v(t,x)) \\ \qquad + \sum_k\left(\sum_i g_{ki}(t,x)\frac{\partial v}{\partial x_i}(t,x) + h_k(t,x;v(t,x))\right)\frac{dW^k}{dt}(t) \\ v(0,x) = v_0(x). \end{cases}$$

Theorem 2.26. *Assume that $u_0(x) \leq v_0(x)$, x a.e., that $f(t,x;r) \leq F(t,x;r)$, t, x a.e., and moreover one of the two pairs $(f,(h_k))$ or $(F,(h_k))$ satisfies the above conditions for existence-uniqueness. Then $u(t,x) \leq v(t,x)$ x a.e., $\mathbb{P}$ a.s., for all $t \geq 0$.*

SKETCH OF THE PROOF OF THEOREM 2.26: The proof is similar to that of Theorem 2.24, so we just sketch it. We first replace v by $u \vee v$ in the last equation, in the three places where we changed u into u^+ in the proof of the previous theorem. The fact that

$$u, v \in H^1(\mathbb{R}^d) \Rightarrow u \vee v \in H^1(\mathbb{R}^d)$$

follows from Lemma 2.25 and the simple identity $u \vee v = u + (v-u)^+$. If v denotes the solution of this new equation, we can show (with the same functional Φ as in the proof of Theorem 2.24) that $\mathbb{E}\Phi(v(t) - u(t)) \leq 0$, which implies that $u(t,x) \leq v(t,x)$, x a.e., $\mathbb{P}$ a.s., for all $t \geq 0$. Consequently v solves the original equation, and the result is established. $\square$

Chapter 3
SPDEs Driven By Space-Time White Noise

3.1 Introduction

The results of the previous chapter mainly apply to equations driven by finite-dimensional Brownian motion or space-time noise which is white in time and colored in space. The aim of this chapter is to study equations driven by space-time white noise. We first explain why we need to restrict ourselves to the case of a one-dimensional space variable. Then we present the basic existence-uniqueness result, together with the Hölder continuity of the solution, following the by now classical St Flour notes of J. Walsh [30].

We will then give sufficient conditions for the solution to be nonnegative. Next we present the application of Malliavin calculus to white noise-driven SPDEs, which allows us to give sufficient conditions for the law of the random variable $u(t, x)$ (or of the random vector $(u(t, x_1), \ldots, u(t, x_n))$) to have a density w.r.t. Lebesgue measure. We next discuss the connection between SPDEs and the super Brownian motion, and finally reflected SPDEs.

3.2 Restriction to a One-Dimensional Space Variable

Let us consider the following linear parabolic SPDE

$$\begin{cases} \dfrac{\partial u}{\partial t}(t, x) = \dfrac{1}{2}\Delta u(t, x) + \mathring{W}(t, x), \ t \geq 0, x \in \mathbb{R}^d \\ u(0, x) = u_0(x), \quad x \in \mathbb{R}^d. \end{cases}$$

The driving noise in this equation is the so-called "space-time white noise", that is, $\mathring{W}$ is a generalized centered Gaussian field, with covariance given by

$$\mathbb{E}[\mathring{W}(h)\mathring{W}(k)] = \int_0^\infty \int_{\mathbb{R}^d} h(t, x)k(t, x)\mathrm{d}x\mathrm{d}t, \ \forall h, k \in L^2(\mathbb{R}_+ \times \mathbb{R}^d).$$

Since the equation is linear, that is, the mapping

$$\mathring{W} \to u$$

is affine, it always has a solution as a distribution, the driving noise being a random distribution. But we want to know when that solution is a standard stochastic process $\{u(t,x),\ t \geq 0, x \in \mathbb{R}^d\}$. Let

$$p(t,x) = \frac{1}{(2\pi t)^{d/2}} \exp\left(-\frac{|x|^2}{2t}\right).$$

The solution of the above equation is given by

$$u(t,x) = \int_{\mathbb{R}^d} p(t,x-y)u_0(y)dy + \int_0^t \int_{\mathbb{R}^d} p(t-s,x-y)W(ds,dy),$$

at least if the second integral makes sense. Since it is a Wiener integral, it is a centered Gaussian random variable, and we just have to check that its variance is finite. But that variance equals

$$\int_0^t \int_{\mathbb{R}^d} p^2(t-s,x-y)dyds = \frac{1}{(2\pi)^d} \int_0^t \frac{ds}{(t-s)^d} \int_{\mathbb{R}^d} \exp\left(-\frac{|x-y|^2}{t-s}\right)dy$$

$$= \frac{1}{2^d \pi^{d/2}} \int_0^t \frac{ds}{(t-s)^{d/2}} < \infty$$

if and only if $d = 1$! When $d \geq 2$, the solution is a generalized stochastic process, given by

$$(u(t), \varphi) = \int_{\mathbb{R}^d} \int_{\mathbb{R}^d} \varphi(x)p(t,x-y)u_0(y)dxdy$$

$$+ \int_0^t \int_{\mathbb{R}^d} \left(\int_{\mathbb{R}^d} \varphi(x)p(t-s,x-y)dx\right) W(ds,dy), \quad t \geq 0, \varphi \in C_C^\infty(\mathbb{R}^d).$$

Here the second integral is well defined. Indeed, let us assume that $\mathrm{supp}\varphi \subset \overline{B}(0,r)$. Then

$$\int_{\mathbb{R}^d} \varphi(x)p(t-s,x-y)dx = \mathbb{E}_y\varphi(B_{t-s}),$$

where $\{B_t,\ t \geq 0\}$ is a standard $\mathbb{R}^d$-valued Brownian motion. For $|y| > 2r$,

$$|\mathbb{E}_y\varphi(B_{t-s})| = |\mathbb{E}_y\left[\varphi(B_{t-s})\mathbf{1}_{|B_{t-s}|\leq r}\right]|$$

$$\leq \|\varphi\|_\infty \mathbb{P}_0(|B_{t-s}| \geq |y| - r)$$

$$\leq \|\varphi\|_\infty \frac{\mathbb{E}_0(|B_{t-s}|^p)}{(|y|-r)^p}$$

$$\leq c(d,p)\|\varphi\|_\infty \frac{(t-s)^{p/2}}{(|y|-r)^p}.$$

Choosing $p > d$, we conclude that

$$\int_0^t \int_{\mathbb{R}^d} \left(\int_{\mathbb{R}^d} \varphi(x) p(t-s, x-y) dx \right)^2 ds dy < \infty.$$

We note that our goal is to solve nonlinear equations of the type

$$\begin{cases} \frac{\partial u}{\partial t}(t,x) = \frac{1}{2} \Delta u(t,x) + f(u(t,x)) + g(u(t,x)) \mathring{W}(t,x), \ t \geq 0, x \in \mathbb{R}^d \\ u(0,x) = u_0(x), \quad x \in \mathbb{R}^d, \end{cases}$$

whose solution might not be more regular than that of the linear equation we considered above. Since we do not want to define the image by a nonlinear mapping of a distribution (which is essentially impossible, if we want to have some reasonable continuity properties, which is crucial when studying SPDEs), we have to restrict ourselves to the case $d = 1$!

3.3 A General Existence-Uniqueness Result

Let us consider specifically the following SPDE with homogeneous Dirichlet boundary conditions

$$\begin{cases} \frac{\partial u}{\partial t}(t,x) = \frac{\partial^2 u}{\partial x^2}(t,x) + f(t,x; u(t,x)) + g(t,x; u(t,x)) \mathring{W}(t,x), \ t \geq 0, 0 \leq x \leq 1; \\ u(t,0) = u(t,1) = 0, \ t \geq 0; \\ u(0,x) = u_0(x), \ 0 \leq x \leq 1. \end{cases}$$

$$(3.1)$$

This equation turns out not to have a classical solution. So we first introduce a *weak formulation* of (3.1), namely

$$\begin{cases} \int_0^1 u(t,x)\varphi(x)dx = \int_0^1 u_0(x)\varphi(x)dx + \int_0^t \int_0^1 u(s,x)\varphi''(x)dxds \\ + \int_0^t \int_0^1 f(s,x;u(s,x))\varphi(x)dxds + \int_0^t \int_0^1 g(s,x;u(s,x))\varphi(x)W(ds,dx) \\ \mathbb{P} \text{ a.s.,} \quad \forall \varphi \in C^2(0,1) \cap C_0([0,1]), \end{cases}$$

$$(3.2)$$

where $C_0([0,1])$ stands for the set of continuous functions from $[0,1]$ into $\mathbb{R}$, which are 0 at 0 and at 1. We need to define the stochastic integral which appears in (3.2). From now on, $W(ds, dx)$ will be considered as a random Gaussian "measure" (it is in fact not a measure for fixed ω) on $\mathbb{R}_+ \times [0,1]$. More precisely, we define the collection

$$\left\{ \mathring{W}(A) = \int_A W(ds, dx), \ A \text{ Borel subset of } \mathbb{R}_+ \times [0,1] \right\}$$

as a centered Gaussian random field with covariance given by

$$\mathbb{E}[\mathring{W}(A)\mathring{W}(B)] = \lambda(A \cap B),$$

where λ denotes Lebesgue measure on $\mathbb{R}_+ \times [0, 1]$.

We want now to sketch the construction of Itô type stochastic integrals with respect to $W(ds, dx)$, where the integrand is allowed to be random, with a restriction of adaptedness in the s direction, but not in the x direction. We refer to Walsh [30] for a more detailed construction.

We define for each $t > 0$ the σ-algebra

$$\mathcal{F}_t = \sigma\{\mathring{W}(A), \ A \ \text{Borel subset of} \ [0, t] \times [0, 1]\},$$

and the associated σ-algebra of predictable sets defined as

$$\mathcal{P} = \sigma\{(s, t] \times \Lambda \subset \mathbb{R}_+ \times \Omega : 0 \le s \le t, \Lambda \in \mathcal{F}_s\}.$$

The class of processes which we intend to integrate with respect to the above measure is the set of functions

$$\psi : \mathbb{R}_+ \times [0, 1] \times \Omega \to \mathbb{R},$$

which are $\mathcal{P} \otimes \mathcal{B}([0, 1])$-measurable and such that

$$\int_0^t \int_0^1 \psi^2(s, x)dxds < \infty \quad \mathbb{P} \ \text{a.s.} \ \forall t \ge 0.$$

In fact we could integrate progressively measurable (and not necessarily predictable) processes. For such ψ's, the stochastic integral

$$\int_0^t \int_0^1 \psi(s, x)W(ds, dx), \quad t \ge 0$$

can be constructed as the limit in probability of the sequence of approximations

$$\sum_{i=1}^{\infty} \sum_{j=0}^{n-1} (\psi, \mathbf{1}_{A_{i-1,j}^n})_{L^2(\mathbb{R}_+ \times (0,1))} W\left(A_{i,j}^n \cap ([0, t] \times [0, 1])\right),$$

where

$$A_{i,j}^n = \left[\frac{i}{n}, \frac{i+1}{n}\right] \times \left[\frac{j}{n}, \frac{j+1}{n}\right].$$

This stochastic integral is a local martingale, with associated increasing process

$$\int_0^t \int_0^1 \psi^2(s, x)dxds, \quad t \ge 0.$$

If, moreover,

$$\mathbb{E} \int_0^t \int_0^1 \psi^2(s, x)dxds, \quad \forall t \ge 0,$$

then the stochastic integral process is a square integrable martingale, the above convergence holds in $L^2(\Omega)$, and we have the isometry

$$\mathbb{E}\left[\left(\int_0^t \int_0^1 \psi(s,x)W(ds,dx)\right)^2\right] = \mathbb{E}\int_0^t \int_0^1 \psi^2(s,x)dxds, \ \forall t \geq 0.$$

We introduce another formulation of our white noise-driven SPDE, namely the integral formulation, which is the following

$$
\begin{cases}
u(t,x) = \displaystyle\int_0^1 p(t;x,y)u_0(y)dy + \int_0^t \int_0^1 p(t-s;x,y)f(s,y;u(s,y))dyds \\[2mm]
\quad + \displaystyle\int_0^t \int_0^1 p(t-s;x,y)g(s,y;u(s,y))W(ds,dy), \ \mathbb{P} \text{ a.s. }, t \geq 0, 0 \leq x \leq 1;
\end{cases}
$$

$$(3.3)$$

where $p(t;x,y)$ is the fundamental solution of the heat equation with Dirichlet boundary condition

$$
\begin{cases}
\dfrac{\partial u}{\partial t}(t,x) = \dfrac{\partial^2 u}{\partial x^2}(t,x); \ t \geq 0, \ 0 < x < 1; \\[2mm]
u(t,0) = u(t,1) = 0, \quad t \geq 0;
\end{cases}
$$

and $u_0 \in C_0([0,1])$. We shall admit the following Lemma (see Walsh [30])

Lemma 3.1. *The above kernel is given explicitly by the formula*

$$p(t;x,y) = \frac{1}{\sqrt{4\pi t}} \sum_{n\in\mathbb{Z}} \left[\exp\left(-\frac{(2n+y-x)^2}{4t}\right) - \exp\left(-\frac{(2n+y+x)^2}{4t}\right)\right],$$

and for all $T > 0$, there exists a C_T such that

$$|p(t;x,y)| \leq \frac{C_T}{\sqrt{t}} \exp\left(-\frac{|x-y|^2}{4t}\right), \quad 0 \leq t \leq T, \ 0 \leq x,y \leq 1.$$

Moreover, there exists a smooth function H such that

$$p(t;x,y) = \frac{1}{\sqrt{4\pi t}} \exp\left(-\frac{|x-y|^2}{4t}\right) + H(t;x,y). \qquad (3.4)$$

We now state two assumptions on the coefficients

$$(H1-n) \quad \sup_{0\leq s\leq t} \int_0^1 (f^{2n}(s,x;0) + g^{2n}(s,x;0))dx < \infty, \ t \geq 0.$$

There exists a locally bounded function $\delta : \mathbb{R} \to \mathbb{R}_+$ such that

$$(H2) \quad |f(s,x;r) - f(s,x,0)| + |g(s,x;r) - g(s,x,0)| \leq \delta(r),$$

$$\forall t \geq 0, 0 \leq x \leq 1, r \in \mathbb{R}.$$

We can now establish the following proposition.

Proposition 3.2. *Under the assumptions $(H1-1)$ and $(H2)$, a continuous $\mathcal{P} \otimes \mathcal{B}([0,1])$-measurable function u satisfies (3.2) if and only if it satisfies (3.3).*

PROOF: First let u be a solution of (3.2), and $\lambda \in C^1(\mathbb{R}_+)$. Then by integration by parts (we use $(\cdot, \cdot)$ to denote the scalar product in $L^2(0, 1)$)

$$
\begin{cases}
\lambda(t)(u(t), \varphi) = \lambda(0)(u(0), \varphi) + \int_0^t (u(s), \lambda(s)\varphi'' + \lambda'(s)\varphi)ds \\
+ \int_0^t \lambda(s)(f(s, \cdot; u(s, \cdot)), \varphi)ds + \int_0^t \int_0^1 \lambda(s)g(s, x; u(s, x))\varphi(x)W(ds, dx).
\end{cases}
$$

But any $\phi \in C^{1,2}(\mathbb{R}_+ \times (0, 1)) \cap C(\mathbb{R}_+ \times [0, 1])$ such that $\phi(t, 0) = \phi(t, 1) = 0$ is a limit of finite sums of the form $\sum_{i=1}^n \lambda_i(t)\varphi_i(x)$. Hence we get that for all ϕ as above and all $t \geq 0$,

$$
\begin{cases}
(u(t), \phi(t, \cdot)) = (u(0), \phi(0, \cdot)) + \int_0^t \left(u(s), \dfrac{\partial^2 \phi}{\partial x^2}(s, \cdot) + \dfrac{\partial \phi}{\partial s}(s, \cdot)\right)ds \\
+ \int_0^t (f(s, \cdot; u(s, \cdot)), \phi(s, \cdot))ds + \int_0^t \int_0^1 \phi(s, x)g(s, x; u(s, x))W(ds, dx).
\end{cases}
$$

Now, t being fixed, we choose for $0 \leq s \leq t, 0 \leq x \leq 1$,

$$
\phi(s, x) = \int_0^1 p(t - s; y, x)\varphi(y)dy = p(t - s; \varphi, x),
$$

where $\varphi \in C_0^\infty([0, 1])$. We deduce that

$$
\begin{cases}
(u(t), \varphi) = (u(0), p(t; \varphi, \cdot)) + \int_0^t (f(s, \cdot; u(s, \cdot)), p(t - s; \varphi, \cdot))ds \\
+ \int_0^t \int_0^1 p(t - s; \varphi, y)g(s, y; u(s, y))W(ds, dy).
\end{cases}
$$

If we now let φ tend to δ_x, we obtain (3.3).

Let now u be a solution of (3.3). Then for all $\varphi \in C^2(0, 1) \cap C_0([0, 1])$, $t \geq 0$, we have, for all $0 \leq s \leq t$,

$$
\begin{cases}
(u(t), \varphi) = (u(s), p(t - s, \varphi, \cdot)) + \int_s^t (f(r, \cdot; u(r, \cdot)), p(t - r; \varphi, \cdot))ds \\
+ \int_s^t \int_0^1 p(t - r; \varphi, y)g(r, y; u(r, y))W(dr, dy).
\end{cases}
$$

We next define $t_i = it/n$, for $0 \leq i \leq n$, and $\Delta t = t/n$.

$$
u(t, \varphi) - (u_0, \varphi) = \sum_{i=0}^{n-1} [(u(t_{i+1}), \varphi) - (u(t_i), \varphi)]
$$

$$
= \sum_{i=0}^{n-1} [(u(t_{i+1}), \varphi) - (u(t_i), p(\Delta t, \varphi, \cdot)) + (u(t_i), p(\Delta t, \varphi, \cdot)) - (u(t_i), \varphi)]
$$

$$= \sum_{i=0}^{n-1} \left[\int_{t_i}^{t_{i+1}} \int_0^1 p(t_{i+1} - s, \varphi, y) f(s, y; u(s, y)) dy ds \right.$$

$$+ \int_{t_i}^{t_{i+1}} \int_0^1 p(t_{i+1} - s, \varphi, y) g(s, y; u(s, y)) W(dy, ds)$$

$$\left. + \int_{t_i}^{t_{i+1}} \int_0^1 u(t_i, y) \frac{\partial^2 p}{\partial y^2}(s - t_i, \varphi, y) dy ds \right]$$

If we exploit the fact that u is a.s. continuous and adapted, we obtain that as $n \to \infty$, the last expression tends to

$$\int_0^t \int_0^1 \varphi(y) f(s, y; u(s, y)) dy ds + \int_0^t \int_0^1 \varphi(y) g(s, y; u(s, y)) W(dy, ds)$$

$$+ \int_0^t \int_0^1 u(s, y) \varphi''(y) dy ds. \qquad \square$$

In order to prove the existence and uniqueness of a solution, we need to replace the assumption $(H2)$ by the stronger assumption: there exists a $k > 0$ such that for all t, x, r, r',

$$(H3) \quad |f(t, x, r) - f(t, x, r')| + |g(t, x, r) - g(t, x, r')| \le k|r - r'|.$$

We have the following theorem.

Theorem 3.3. *Under the assumptions $(H1 - n)$ for all $n \ge 1$ and $(H3)$, if $u_0 \in C_0([0, 1])$, there exists a unique continuous $\mathcal{P} \otimes \mathcal{B}([0, 1])$-measurable solution u of equation (3.3). Moreover, $\sup_{0 \le x \le 1,\, 0 \le t \le T} \mathbb{E}[|u(t, x)|^p] < \infty$, for all $p \ge 1$.*

PROOF: UNIQUENESS Let u and v be two solutions. Then the difference $\bar{u} = u - v$ satisfies

$$\bar{u}(t, x) = \int_0^t \int_0^1 p(t - s; x, y)[f(s, y; u(s, y)) - f(s, y; v(s, y))] ds dy$$

$$+ \int_0^t \int_0^1 p(t - s; x, y)[g(s, y; u(s, y)) - g(s, y; v(s, y))] W(ds, dy).$$

Using successively the inequality $(a + b)^2 \le 2(a^2 + b^2)$, Cauchy–Schwarz, the isometry property of the stochastic integral, and $(H3)$, we obtain

$$\mathbb{E}[\bar{u}^2(t, x)] \le 2(t + 1)k^2 \int_0^t \int_0^x p^2(t - s; x, y) \mathbb{E}[\bar{u}^2(s, y)] dy ds$$

Let $H(t) = \sup_{0 \le x \le 1} \mathbb{E}[\bar{u}^2(t, x)]$. We deduce from the last inequality

$$H(t) \le 2(t + 1) \int_0^t \left[\sup_{0 \le x \le 1} \int_0^1 p^2(t - s; x, y) dy \right] H(s) ds.$$

From the above estimate on p, we deduce that

$$\sup_{0\le x\le 1}\int_0^1 p^2(t-s;x,y)dy \le \frac{C_T^2}{t-s}\int_{\mathbb{R}}\exp\left(-\frac{|x-y|^2}{2(t-s)}\right)dy \le \frac{C'}{\sqrt{t-s}},$$

and iterating twice the estimate thus obtained for H, we deduce that

$$H(t) \le C''\int_0^t H(s)ds,$$

hence $H(t) = 0$ from Gronwall's Lemma.

EXISTENCE We use the well-known Picard iteration procedure

$$u^0(t,x) = 0$$

$$u^{n+1}(t,x) = \int_0^1 p(t;x,y)u_0(y)dy + \int_0^t\int_0^1 p(t-s;x,y)f(s,y;u^n(s,y))dyds$$

$$+ \int_0^t\int_0^1 p(t-s;x,y)g(s,y;u^n(s,y))W(dy,ds).$$

Let $H_n(t) = \sup_{0\le x\le 1}\mathbb{E}[|u^{n+1}(t,x)-u^n(t,x)|^2]$. Then, as in the proof of uniqueness, we have that for $0 \le t \le T$,

$$H_n(t) \le C_T\int_0^t H_{n-2}(s)ds.$$

Iterating this inequality k times, we get

$$H_n(t) \le C_T^k\int_0^t ds_1\int_0^{s_1} ds_2\cdots\int_0^{s_{k-1}} H_{n-2k}(s_k)ds_k$$

$$\le \frac{C_T^k t^{k-1}}{(k-1)!}\int_0^t ds H_{n-2k}(s).$$

But

$$H_0(t) = \sup_{0\le x\le 1}\mathbb{E}\left(\left|\int_0^1 p(t;x,y)u_0(y)dy + \int_0^t\int_0^1 p(t-s;x,y)f(s,y;0)dyds\right.\right.$$

$$\left.\left.+ \int_0^t\int_0^1 p(t-s;x,y)g(s,y;0)W(dy,ds)\right|^2\right) < \infty,$$

thanks to assumption $(H1)$. Hence the sequence $\{u^n\}$ is Cauchy in $L^\infty((0,T)\times(0,1);L^2(\Omega))$; its limit u is $\mathcal{P}\times\mathcal{B}([0,1])$-measurable and satisfies (3.3). All of the arguments are valid with the exponent 2 replaced by p, hence the p-th moment estimate. It remains to show that it can be taken to be continuous, which we will do in the next theorem.

Theorem 3.4. *The solution u of equation (3.3) has a modification which is a.s. Hölder continuous in (t,x), with the exponent $1/4 - \varepsilon$, $\forall \varepsilon > 0$.*

In fact u is $1/4 - \varepsilon$–Hölder continuous in t, and $1/2 - \varepsilon$–Hölder continuous in x.

PROOF: It suffices to show that each term in the right-hand side of (3.3) has the required property. We shall only consider the stochastic integral term, which is the hardest. Consider

$$v(t,x) = \int_0^t \int_0^1 p(t-s;x,y)g(s,y;u(s,y))W(ds,dy).$$

We use the following well-known Kolmogorov Lemma

Lemma 3.5. *If $\{X_\alpha, \; \alpha \in D \subset \mathbb{R}^d\}$ is a random field such that for some k, n and $\beta > 0$, for all $\alpha, \alpha' \in D$,*

$$\mathbb{E}\left(|X_\alpha - X_{\alpha'}|^n\right) \le k|\alpha - \alpha'|^{d+\beta},$$

then there exists a modification of the process $\{X_\alpha\}$ which is a.s. Hölder continuous with the exponent $\beta/n - \varepsilon$, for all $\varepsilon > 0$.

PROOF OF THEOREM 3.4 We follow Walsh [30]. It is not hard to show that it suffices to prove the needed estimates with $p(t-s;x,y)$ replaced by the first term on the right of (3.4). Hence in this proof we proceed as if the second term on the right of (3.4) is zero. We first note that

$$\mathbb{E}[|(v(t+k,x+h) - v(t,x)|^n]^{1/n} \le \mathbb{E}[|(v(t+k,x+h) - v(t+k,x)|^n]^{1/n}$$
$$+ \mathbb{E}[|(v(t+k,x) - v(t,x)|^n]^{1/n}.$$

We estimate the first term (for simplicity of notation, we replace $t + k$ by t). From Burkholder and Hölder,

$$\mathbb{E}[|(v(t,x+h) - v(t,x)|^n]$$

$$\le c\mathbb{E}\left(\left|\int_0^t \int_0^1 g^2(u;s,y)[p(t-s;x+h,y) - p(t-s;x,y)]^2 dyds\right|^{n/2}\right)$$

$$\le c\mathbb{E}\left(\int_0^t \int_0^1 g^n(u;s,y)dsdy\right)$$

$$\times \left(\int_0^t \int_{-\infty}^\infty |p(s;x,z) - p(s;x+h,z)|^{2n/(n-2)}dzds\right)^{(n-2)/2}.$$

The first factor on the right is bounded by a constant depending upon n only, thanks to Assumption $(H1 - n/2)$ and the estimate obtained in Theorem 3.3.

We next consider the second factor in the above right-hand side. We have, with $h = hz$, $s = h^2 v$,

$$\int_0^t \int_{-\infty}^\infty |p(s;x,z) - p(s;y,z)|^{2n/(n-2)}dzds$$

$$= ch^{(n-6)/(n-2)} \int_0^t \int_{-\infty}^\infty v^{-n/(n-2)} \left|e^{-\frac{|z+1|^2}{4v}} - e^{-\frac{|z|^2}{4v}}\right|^{2n/(n-2)} dvdz$$

$$= Ch^{(n-6)/(n-2)},$$

provided the integral converges, which is the case whenever $n > 6$. In this case, we have proved that

$$\mathbb{E}[|(v(t, x + h) - v(t, x)|^n] \le C_n |h|^{(n-6)/2},$$

and $x \to v(t, x)$ is Hölder with any exponent $< 1/2$.

Analogously

$$\mathbb{E}[|(v(t + k, x) - v(t, x)|^n]$$

$$\le c\mathbb{E}\left(\left|\int_0^t \int_0^1 g^2(u; s, y)[p(t + k - s; x, y) - p(t - s; x, y)]^2 dy ds\right|^{n/2}\right) \quad (3.5)$$

$$+ c\mathbb{E}\left(\left|\int_t^{t+k} \int_0^1 g^2(u; s, y) p^2(t + k - s; x, y) dy ds\right|^{n/2}\right).$$

The first term on the right of (3.5) can be estimated as follows.

$$\mathbb{E}\left(\left|\int_0^t \int_0^1 g^2(u; s, y)[p(t + k - s; x, y) - p(t - s; x, y)]^2 dy ds\right|^{n/2}\right)$$

$$\le c\mathbb{E}\left(\int_0^t \int_0^1 g^n(u; s, y) ds dy\right)$$

$$\times \left(\int_0^t \int_{-\infty}^\infty |p(t + k - s; x, y) - p(t - s; x, y|^{2n/(n-2)} dy ds\right)^{(n-2)/2}$$

$$\le C_n \left(\int_0^t \int_{-\infty}^\infty |p(s + k; 0, y) - p(s; 0, y|^{2n/(n-2)} dy ds\right)^{(n-2)/2}$$

$$= C_n \left(k^{\frac{3}{2} - \frac{n}{n-2}} \int_0^t \int_{-\infty}^\infty \left(\frac{e^{-\frac{z^2}{4(u+1)}}}{\sqrt{u+1}} - \frac{e^{-\frac{z^2}{4u}}}{\sqrt{u}}\right)^{\frac{2n}{n-2}} dz du\right)^{(n-2)/2}$$

$$= C_n' k^{\frac{n}{4} - \frac{3}{2}},$$

where we have defined $u = s/k$, $z = y/\sqrt{k}$, and $C_n' < \infty$ provided $n > 6$.

We finally estimate the second term on the right of (3.5). It is bounded by a constant times (in the following computation, the value of the constant C changes from line to line)

$$\mathbb{E}\int_t^{t+k} \int_0^1 g^n(u; s, y) dy ds \times \left(\int_0^k \int_{-\infty}^\infty p^{\frac{2n}{n-2}}(s; 0, y) dy ds\right)^{\frac{n-2}{n}}$$

$$\le Ck \left(\int_0^k \int_{-\infty}^\infty s^{-\frac{n}{n-2}} e^{-\frac{2n}{n-2}\frac{y^2}{4s}} dy ds\right)^{\frac{n-2}{n}}$$

$$= Ck \left(\int_0^k s^{-\frac{1}{2} - \frac{2}{n-2}} ds\right)^{\frac{n-2}{n}}$$

$$= Ck^{\frac{n}{4}-\frac{1}{2}}.$$

Hence $t \to v(t,x)$ is a.s. Hölder with any exponent $< 1/4$. 					□

3.4 A More General Existence and Uniqueness Result

One can generalize the existence-uniqueness result to coefficients satisfying the following assumptions (see Zangeneh [32] and Gyöngy, Pardoux [9])

$$(A1) \begin{cases} \forall T, R, \quad \exists K(T, R) \text{ such that } \forall 0 \le x \le 1, t \le T, |r|, |r'| \le R \\ (r - r')[f(t,x;r) - f(t,x;r')] + |g(t,x;r) - g(t,x;r')|^2 \le K(T, R)|r - r'|^2 \end{cases}$$

$$(A2) \begin{cases} \exists C \text{ such that } \forall t \ge 0, r \in \mathbb{R}, 0 \le x \le 1, \\ rf(t,x;r) + |g(t,x;r)|^2 \le C(1 + |r|^2) \end{cases}$$

$$(A3) \quad \forall t \ge 0,, 0 \le x \le 1, \ r \to f(t,x;r) \text{ is continuous.}$$

Moreover, without the assumption $(A2)$, the solution exists and is unique up to some (possibly infinite) stopping time.

If one suppresses the above condition $(A1)$, and adds the condition that

$$\forall t \ge 0, \ 0 \le x \le 1, \quad r \to g(t,x;r) \text{ is continuous,}$$

then one can show the existence of a weak solution (i.e. a solution of the associated martingale problem).

3.5 Positivity of the Solution

Theorem 3.6. *Let u and v be the two solutions of the two white noise-driven SPDEs*

$$\begin{cases} \dfrac{\partial u}{\partial t}(t,x) = \dfrac{\partial^2 u}{\partial x^2}(t,x) + f(t,x;u(t,x)) + g(t,x;u(t,x))\mathring{W}(t,x), \ t \ge 0, 0 \le x \le 1; \\ u(t,0) = u(t,1) = 0, \ t \ge 0; \\ u(0,x) = u_0(x), \ 0 \le x \le 1. \end{cases}$$

$$(3.6)$$

$$\begin{cases} \dfrac{\partial v}{\partial t}(t,x) = \dfrac{\partial^2 v}{\partial x^2}(t,x) + F(t,x;v(t,x)) + g(t,x;v(t,x))\mathring{W}(t,x), \ t \ge 0, 0 \le x \le 1; \\ v(t,0) = v(t,1) = 0, \ t \ge 0; \\ v(0,x) = v_0(x), \ 0 \le x \le 1. \end{cases}$$

$$(3.7)$$

Assume that u_0, $v_0 \in C_0([0, 1])$ and the two pairs (f, g) or (F, g) satisfy the conditions for strong existence and uniqueness. Then if $u_0(x) \leq v_0(x) \; \forall x$ and $f \leq F$, $u(t, x) \leq v(t, x) \; \forall t \geq 0$, $0 \leq x \leq 1$, $\mathbb{P}$ a.s.

PROOF: Let $\{e_k, \; k \geq 1\}$ be an orthonormal basis of $L^2(0, 1)$. Formally,

$$\mathring{W}(t, x) = \sum_{k=1}^{\infty} \mathring{W}^k(t) e_k(x),$$

where $\mathring{W}^k(t) = (\mathring{W}(t, \cdot), e_k)$. For each $N \geq 1$, let

$$\mathring{W}_N(t, x) = \sum_{k=1}^{N} \mathring{W}^k(t) e_k(x),$$

and u_N (resp. v_N) be the solution of (3.6) (resp. (3.7)), where $\mathring{W}$ has been replaced by $\mathring{W}_N$. We now prove

Lemma 3.7. *For all $n \geq 1$, $T \geq 0$,*

$$\lim_{N \to \infty} \sup_{0 \leq t \leq T, \; 0 \leq x \leq 1} \mathbb{E}[|(u(t, x) - u_N(t, x)|^n] = 0,$$

and the same is true for the difference $v - v_N$.

PROOF: We follow the argument from Lemma 2.1 in [5]. We have the decomposition (we write $f(u)$ and $g(u)$ instead of $f(t, x; u)$ and $g(t, x; u)$)

$$u(t, x) - u_N(t, x) = A_N(t, x) + B_N(t, x) + C_N(t, x), \quad \text{where}$$

$$A_N(t, x) = \int_0^t \int_0^1 [f(u(s, y)) - f(u_N(s, y))] p(t - s; x, y) dy ds,$$

$$B_N(t, x) = \sum_{k=1}^{N} \int_0^t \int_0^1 [g(u(s, y)) - g(u_N(s, y))] p(t - s; x, y) e_k(y) dy dW^k(s),$$

$$C_N(t, x) = \int_0^t \int_0^1 [\Psi(t, x; s, y) - \Psi_N(t, x; s, y)] W(ds, dy), \quad \text{with}$$

$$\Psi(t, x; s, y) = g(u(s, y)) p(t - s; x, y),$$

$$\Psi_N(t, x; s, y) = \sum_{k=1}^{N} \int_0^1 g(u(s, z)) p(t - s; x, z) e_k(z) dz \, e_k(y).$$

We shall use the following property of the kernel $p(t; x, y)$, see Walsh [30]: if $0 < r < 3$, for all $T > 0$,

$$\sup_{0 \leq x \leq 1} \int_0^T \int_0^1 p^r(t; x, y) dy dt < \infty. \tag{3.8}$$

We shall also assume that f and g are globally Lipschitz, which we can do by a localization argument. In the sequel, n will be an exponent satisfying $n > 6$. Then the conjugate exponent $m = \frac{n}{n-1} < 3$. Below C_n will denote a constant which depends only upon n, and may change from line to line. We set

$$F_N(t) = \sup_{0 \le x \le 1} \mathbb{E}[|u(t,x) - u_N(t,x)|^n].$$

Form Hölder's inequality and (3.8),

$$\mathbb{E}[|A_N(t,x)|^n] \le C_n \left(\int_0^t \int_0^1 p^m(s; x, y) dy ds \right)^{n/m}$$

$$\times \mathbb{E} \int_0^1 |u(s,y) - u_N(s,y)|^n dy ds$$

$$\le C_n \int_0^t F_N(s) ds. \tag{3.9}$$

Next

$$\mathbb{E}[|B_N(t,x)|^n] \le$$

$$C\mathbb{E}\left[\left\{ \sum_{k=1}^N \int_0^t \left(\int_0^1 [g(u(s,y)) - g(u_N(s,y))] p(t-s; x, y) e_k(y) dy \right)^2 ds \right\}^{n/2} \right].$$

But, denoting $L^2(0,1)$ by H,

$$\sum_{k=1}^N \left(\int_0^1 [g(u(s,y)) - g(u_N(s,y))] p(t-s; x, y) e_k(y) dy \right)^2$$

$$= \sum_{k=1}^N ([g(u(s,\cdot)) - g(u_N(s,\cdot))] p(t-s; x, \cdot), e_k)_H^2$$

$$\le \|[g(u(s,\cdot)) - g(u_N(s,\cdot))] p(t-s; x, \cdot)\|_H^2.$$

Consequently, with $m' = \frac{n/2}{n/2-1}$, noting that $2m' < 3$,

$$\mathbb{E}[|B_N(t,x)|^n] \tag{3.10}$$

$$\le C_n \mathbb{E}\left[\left\{ \int_0^t \int_0^1 [g(u(s,y)) - g(u_N(s,y))]^2 p^2(t-s; x, y) dy ds \right\}^{n/2} \right]$$

$$\le C_n \left(\int_0^t \int_0^1 p^{2m'}(s; x, y) dy ds \right)^{1/2m'} \mathbb{E} \int_0^t \int_0^1 |g(u(s,y)) - g(u_N(s,y))|^n dy ds$$

$$\le C_n \int_0^t F_N(s) ds. \tag{3.11}$$

Finally

$$\mathbb{E}[|C_N(t,x)|^n] \le C_n \mathbb{E}\left[\left\{\int_0^t \int_0^1 [\Psi(t,x;s,y) - \Psi_N(t,x;s,y)]^2 dyds\right\}^{n/2}\right].$$

We note that

$$\Psi_N(t,x;s,y) = \sum_{k=1}^N (\Psi(t,x;s,\cdot),e_k)e_k(y), \quad \text{hence}$$

$$\int_0^1 [\Psi(t,x;s,y) - \Psi_N(t,x;s,y)]^2 dy = \|\Psi(t,x;s,\cdot) - \Psi_N(t,x;s,\cdot))\|_H^2$$

$$\downarrow 0 \text{ a.s., as } N \to \infty.$$

Moreover,

$$\|\Psi(t,x;s,\cdot) - \Psi_N(t,x;s,\cdot))\|_H^2 \le \|\Psi(t,x;s,\cdot)\|_H^2, \quad \text{and}$$

$$\mathbb{E}\left[\left\{\int_0^t \|\Psi(t,x;s,\cdot)\|_H^2 ds\right\}^{n/2}\right] < \infty.$$

Hence by the dominated convergence theorem, $\mathbb{E}[|C_N(t,x)|^n] \to 0$, as $N \to \infty$.
Set

$$\gamma_N(t,x) := \mathbb{E}\left[\left\{\int_0^t \int_0^1 [\Psi(t,x;s,y) - \Psi_N(t,x;s,y)]^2 dyds\right\}^{n/2}\right].$$

This is a sequence of continuous functions on the compact set $[0,T] \times [0,1]$, which decreases pointwise to 0, as $N \to \infty$. Hence, by Dini's theorem,[1] γ_N converges uniformly to 0, and $\sup_{0\le x\le 1} \mathbb{E}[|C_N(t,x)|^n] \to 0$, as $N \to \infty$. For any $\varepsilon > 0$, there exists an $N_\varepsilon \ge 1$ such that, if $N \ge N_\varepsilon$, $\sup_{0\le x\le 1} \mathbb{E}[|C_N(t,x)|^n] \le \varepsilon$. From this, combined with (3.9) and (3.10), we deduce that

$$F_N(t) \le C_n \int_0^t F_N(s)ds + C_n\varepsilon.$$

The result now follows from Gronwall's Lemma.   □

We now conclude the proof of Theorem 3.6. From Theorem 2.26, $\mathbb{P}$ a.s. $u_N(t,x) \le v_N(t,x)$ for all $t \ge 0, x \in [0,1], N \ge 1$. Theorem 3.6 then follows, since $u_N(t,x) \to u(t,x)$ and $v_N(t,x) \to u(t,x)$.   □

Corollary 3.8. *Let $u_0(x) \ge 0$, assume (f,g) satisfies the conditions for strong existence-uniqueness of a solution u to equation (3.3). If, moreover,*

$$f(t,x;0) \ge 0, \quad g(t,x;0) = 0, \quad \forall t \ge 0, 0 \le x \le 1,$$

then $u(t,x) \ge 0, \forall t \ge 0, 0 \le x \le 1, \mathbb{P}$ a.s.

---

[1] see e.g. https://en.wikipedia.org/wiki/Dini's_theorem

PROOF: Let $v_0 \equiv 0 \leq u_0(x)$, $F(t,x;r) = f(t,x;r) - f(t,x;0) \leq f(t,x;r)$. Then $v \equiv 0$ solves (3.7), and the result follows from the comparison theorem (reversing the orders). □

3.6 Applications of Malliavin Calculus to SPDEs

We consider again equation (3.3). Our assumptions in this section are the following

$(M1)$ $\begin{cases} \forall 0 \leq x \leq 1, t \geq 0, \quad r \to (f(t,x;r), g(t,x;r)) \text{ is of class } C^1, \\ \text{uniformly in } t \text{ and } x, \text{ and the derivatives are locally bounded.} \end{cases}$

$(M2)$ $\begin{cases} \exists C \text{ such that } \forall t \geq 0, r \in \mathbb{R}, 0 \leq x \leq 1, \\ rf(t,x;r) + |g(t,x;r)|^2 \leq C(1+|r|^2) \end{cases}$

$(M3)$ $\quad \exists y \in (0,1)$ such that $g(0,y;u_0(y)) \neq 0$.

The aim of this section is to show the following result from [27].

Theorem 3.9. *Under conditions $(M1)$, $(M2)$ and $(M3)$, for any $t > 0$, $0 < x < 1$, the law of the random variable $u(t,x)$ is absolutely continuous with respect to Lebesgue measure on $\mathbb{R}$.*

Let us first state and prove one corollary to this result

Corollary 3.10. *Under the conditions of Theorem 3.9, if moreover $u_0(x) \geq 0$, $u_0 \not\equiv 0$, $f(t,x;0) \geq 0$, $g(t,x;0) = 0$, then $u(t,x) > 0$, $\forall t > 0$, x a.e., $\mathbb{P}$ a.s.*

PROOF: From Corollary 3.8, we already know that $u(t,x) \geq 0$ for all t, x, $\mathbb{P}$ a.s. Moreover $\mathbb{P}(u(t,x) = 0) = 0$, hence for each fixed (t,x), $u(t,x) > 0$ $\mathbb{P}$ a.s. The result follows from the continuity of u. □

Let us recall the basic ideas of Malliavin calculus, adapted to our situation. We consider functionals of the Gaussian random measure $\mathring{W}$. We first consider the so-called *simple* random variables, which are of the following form:

$$F = f(\mathring{W}(k_1), \ldots, \mathring{W}(k_n)),$$

where $f \in C_b^\infty(\mathbb{R}^n)$, $k_1, \ldots, k_n \in H = L^2(\mathbb{R}_+ \times (0,1))$. For any $h \in H$, we define the Malliavin derivative of F in the direction h as

$$D_h F = \frac{d}{d\varepsilon} f(\mathring{W}(k_1) + \varepsilon(h,k_1), \ldots, \mathring{W}(k_n) + \varepsilon(h,k_n))|_{\varepsilon=0}$$

$$= \sum_{i=1}^{n} \frac{\partial f}{\partial x_i}(\mathring{W}(k_1), \ldots, \mathring{W}(k_n))(h,k_i),$$

and the first-order Malliavin derivative of F as the random element of H $v(t,x) = D_{tx}F$ given as

$$D_{tx}F = \sum_{i=1}^{n} \frac{\partial f}{\partial x_i}(\mathring{W}(k_1), \ldots, \mathring{W}(k_n))k_i(t,x).$$

Note that

$$D_h F = \int_0^\infty \int_0^1 D_{t,x}Fh(t,x)\mathrm{d}x\mathrm{d}t,$$

which makes sense with h random. In the proof of Theorem 3.9 below, this will allow us to write $D_h u(t,x)$ with h random.

We next define the $\|\cdot\|_{1,2}$ norm of a simple random variable as follows

$$\|F\|_{1,2}^2 = \mathbb{E}(F^2) + \mathbb{E}(|DF|_H^2).$$

Now the Sobolev space $\mathbb{D}^{1,2}$ is defined as the closure of the set of simple random variables with respect to the $\|\cdot\|_{1,2}$ norm. Both the directional derivative D_h and the derivation D are closed operators, which can be extended to elements of the space $\mathbb{D}^{1,2}$. They can even be extended to elements of $\mathbb{D}_{loc}^{1,2}$, which is defined as follows. $X \in \mathbb{D}_{loc}^{1,2}$ whenever there exists a sequence $\{X_n, \ n \geq 1\}$ of elements of $\mathbb{D}^{1,2}$ which are such that the sequence $\Omega_n = \{X = X_n\}$ is increasing, and $\mathbb{P}(\Omega \setminus \cup_n \Omega_n) = 0$. We note that for $X \in \mathbb{D}_{loc}^{1,2}$, which is $\mathcal{F}_t$ measurable, $D_{sy}X = 0$ whenever $s > t$. One should think intuitively of the operator D_{sy} as the derivation of a function of $\mathring{W}$ with respect to $\mathring{W}(s,y)$, the white noise at the point (s,y).

We shall also use the space $\mathbb{D}_h$, which is the closure of the set of simple random variables with respect to the norm whose square is defined as

$$\|X\|_h^2 = \mathbb{E}(F^2 + |D_h F|^2).$$

The following is a simple consequence of a well-known result of Bouleau and Hirsch.

Proposition 3.11. *Let $X \in \mathbb{D}_{loc}^{1,2}$. If $\|DX\|_H > 0$ a.s., then the law of the random variable X is absolutely continuous with respect to Lebesgue measure.*

PROOF: We follow the proof in Nualart [21]. It suffices to treat the case where $X \in \mathbb{D}^{1,2}$ and $|X| < 1$ a.s. It now suffices to show that whenever $g : (-1,1) \to [0,1]$ is measurable,

$$\int_{-1}^{1} g(y)\mathrm{d}y = 0 \Rightarrow \mathbb{E}g(X) = 0.$$

There exists a sequence $\{g_n\} \subset C_b^1((-1,1); [0,1])$ which converges to g a.e. both with respect to the law of X and with respect to Lebesgue measure. Define

$$\psi_n(x) = \int_{-1}^{x} g_n(y)\mathrm{d}y, \quad \psi(x) = \int_{-1}^{x} g(y)\mathrm{d}y.$$

Now $\psi_n(X) \in \mathbb{D}^{1,2}$ and $D[\psi_n(X)] = g_n(X)DX$. Moreover, $\psi_n(X) \to \psi(X)$ in $\mathbb{D}^{1,2}$. We observe that $\psi(X) = 0$, and $D[\psi(X)] = g(X)DX$. Finally from the assumption of the proposition it follows that $g(X) = 0$ a.s.   □

PROOF OF THEOREM 3.9: We shall prove that for fixed (t, x), $u(t, x) \in \mathbb{D}_{\text{loc}}^{1,2}$, then compute the directional Malliavin derivative $D_h u(t, x)$, and finally prove that $\|Du(t, x)\|_H > 0$ a.s.

STEP 1. By the localization argument, it suffices to prove that whenever f, f'_r, g and g'_r are bounded, $u(t, x) \in \mathbb{D}^{1,2}$. We first show that a directional derivative exists in any direction of the form $h(t, x) = \rho(t)e_\ell(x)$, where $\rho \in L^2(\mathbb{R}_+)$, and e_ℓ is an element of an orthonormal basis of $L^2(0, 1)$. This is done by approximating (3.3) by a sequence of finite-dimensional SDEs indexed by n, driven by a finite-dimensional Wiener process. The derivative of the approximate SDE is known to solve a linearized equation, which converges as $n \to \infty$ to the solution $v(t, x)$ of the linearized equation

$$\begin{cases} \dfrac{\partial v}{\partial t} = \dfrac{\partial^2 v}{\partial x^2} + f'(u)v + g'(u)v\mathring{W} + g(u)h \\ v(0, x) = 0, \end{cases} \tag{3.12}$$

and the fact that D_h is closed (which means that if $\{X_n\} \subset \mathbb{D}_h$, $X_n \to X$ in $L^2(\Omega)$, $D_h X_n \to Y$ in $L^2(\Omega; H)$, then $X \in \mathbb{D}_h$ and $Y = D_h X$) allows us to deduce that $u \in \mathbb{D}_h$ and $v = D_h u$. The fact that $u \in \mathbb{D}^{1,2}$ is proved by showing that, whenever $\{h_n, n \geq 1\}$ is an orthonormal basis of $H = L^2(\mathbb{R}_+ \times (0, 1))$,

$$\mathbb{E}\left(\|Du(t, x)\|_H^2 \right) = \sum_n \mathbb{E}\left(|D_{h_n} u(t, x)|^2 \right),$$

which can be shown to be finite using classical estimates of the kernel of the heat equation.

STEP 2 Let y be such that $g(0, y; u_0(y)) \neq 0$, and suppose for example that $g(0, y; u_0(y)) > 0$. Then there exist $\varepsilon > 0$ and a stopping time τ such that $0 < \tau \leq t$, such that

$$g(s, z; u(s, z)) > 0, \quad \forall z \in [y - \varepsilon, y + \varepsilon], \ 0 \leq s \leq \tau,$$

and we have

$$\|Du(t, x)\|_H > 0 \iff \int_0^t \int_0^1 |D_{s,z} u(t, x)| dz ds > 0.$$

A sufficient condition for this to be true is that

$$\int_0^\tau \int_{y-\varepsilon}^{y+\varepsilon} |D_{s,z} u(t, x)| dz ds > 0.$$

But, $\forall h \in L^2(\Omega \times \mathbb{R}_+ \times [0, 1], \mathcal{P} \otimes \mathcal{B}([0, 1]), \mathbb{P} \times \lambda)$ (λ denoting Lebesgue measure on $[0, +\infty) \times [0, 1]$) such that $h \geq 0$ and $\mathrm{supp}\, h \subset \{(s, y); \ g(s, y; u(s, y)) \geq 0\}$, $D_h u(t, x) \geq 0$, as a consequence of Corollary 3.8, applied to (3.12). Hence a sufficient condition for $\|Du(t, x)\|_H$ to be positive is that

$$\int_0^\tau \int_{y-\varepsilon}^{y+\varepsilon} D_{s,z} u(t, x) dz ds = \int_0^\tau v(s; t, x) ds > 0,$$

where we have defined $v(s; t, x) = \int_{y-\varepsilon}^{y+\varepsilon} D_{s,z} u(t, x) dz$. Let us just show that $v(t, x) = v(0; t, x) > 0$. It is not hard to verify that v solves the linearized SPDE

$$\begin{cases} \dfrac{\partial v}{\partial t} = \dfrac{\partial^2 v}{\partial x^2} + f'(u)v + g'(u)v\mathring{W} \\ v(0, x) = g(0, x; u_0(x))\mathbf{1}_{[y-\varepsilon, y+\varepsilon]}(x). \end{cases} \tag{3.13}$$

Now there exists a $\beta > 0$ such that $g(0, x, u_0(x)) \geq \beta$, for $x \in [y - \varepsilon, y + \varepsilon]$, then by the comparison theorem it suffices to prove our result with the initial condition of (3.13) replaced by $\beta \mathbf{1}_{[y-\varepsilon, y+\varepsilon]}(x)$, and by linearity it suffices to treat the case $\beta = 1$. In order to simplify the notation, we let $a = y - \varepsilon$, and $b = y + \varepsilon$. Since $\bar{v} = e^{ct}v$ satisfies the same equation as v, with $f'(u)$ replaced by $f'(u) + c$, it suffices, again by the comparison theorem, to treat the case $f'(u) \equiv 0$. Finally we need to examine the random variable

$$v(t, x) = v_1(t, x) + v_2(t, x)$$

$$= \int_a^b p(t; x, z) dz + \int_0^t \int_0^1 p(t - s; x, z) g'(u)(s, z) v(s, z) W(ds, dz).$$

Assume that $x \geq a$ (if this is not the case, then we have $x \leq a$, and we can adapt the argument correspondingly). Let d be such that $x \leq b + d < 1$, and define

$$\alpha = \frac{1}{2} \inf_{1 \leq k \leq m} \inf_{a \leq y \leq b + dk/m} \int_a^{b+d(k-1)/m} p\left(\frac{t}{m}; y, z\right) dz.$$

We have that $\alpha > 0$. We now define, for $1 \leq k \leq m$, the event

$$E_k = \left\{ v\left(\frac{kt}{m}, \cdot\right) \geq \alpha^k \mathbf{1}_{[a, b+kd/m]}(\cdot) \right\}.$$

Let us admit for a moment the following lemma.

Lemma 3.12. *For any $\delta > 0$, there exists an $m_\delta \geq 1$ such that for any $m \geq m_\delta$,*

$$\sup_{0 \leq k \leq m-1} \mathbb{P}(E_{k+1}^c | E_1 \cap \cdots \cap E_k) \leq \frac{\delta}{m}.$$

Now

$$\mathbb{P}(v(t, x) > 0) \geq \lim_{m \to \infty} \mathbb{P}(E_1 \cap \cdots \cap E_m) \geq \lim_m \left(1 - \frac{\delta}{m}\right)^m = e^{-\delta},$$

hence the result, since we can let $\delta \to 0$.

PROOF OF LEMMA 3.12: Proving the lemma amounts to proving that $\mathbb{P}(E_1^c) \leq \delta/m$. By the definition of α,

$$v_1\left(\frac{t}{m}, \cdot\right) \geq 2\alpha \mathbf{1}_{[a, b+d/m]}(\cdot).$$

Hence it suffices to show that for any $\delta > 0$, there exists an $m_\delta \geq 1$ such that if $m \geq m_\delta$,

$$\mathbb{P}\left(\sup_{a \leq y \leq b+d/m} \left|v_2\left(\frac{t}{m}, y\right)\right| > \alpha\right) \leq \frac{\delta}{m}.$$

For this to be true, it suffices that there exists $n, p > 1$ and $c > 0$ such that

$$\mathbb{E}\left(\sup_{0 \leq y \leq 1} |v_2(t, y)|^n\right) \leq ct^p.$$

But

$$\mathbb{E}(|v_2(t, y)|^n) \leq c\left(\int_0^t \int_0^1 p^2(t-s; y, z)dzds\right)^{n/2}$$

$$\leq c\left(\int_0^t \int_0^1 p^r(t-s; y, z)dzds\right)^{n/2} t^{n/q},$$

if $\frac{2}{r} + \frac{2}{q} = 1$. Since we need $r < 3$ for the first factor to be finite, we get that for $q > 6$,

$$\mathbb{E}(|v_2(t, y)|^n) \leq ct^{n/q}.$$

Moreover, from the computations in the proof of Theorem 3.4,

$$\mathbb{E}\left(|v_2(t, x) - v_2(t, y)|^n\right) \leq c|x - y|^{\frac{n}{2}-1} t^{n/q}.$$

Choosing $n > q > 6$, this concludes the proof. □

In the case where g does not vanish, and the coefficients are smooth, for any $0 < x_1 \cdots < x_n < 1$, the law of the random vector

$$(u(t, x_1, u(t, x_2), \ldots, u(t, x_n))$$

has a density with respect to Lebesgue measure on $\mathbb{R}^n$, which is everywhere strictly positive. It is an open problem to show the same result under a condition similar to that of Theorem 3.9.

In the case of the 2D Navier–Stokes equation driven by certain low-dimensional white noises, Mattingly and Pardoux [16] have shown that for any $t > 0$, the projection of $u(t, \cdot)$ on any finite-dimensional subspace has a density with respect to Lebesgue measure, which under appropriate conditions is smooth and everywhere positive.

3.7 SPDEs and the Super Brownian Motion

In this section, we want to study the SPDE

$$\begin{cases} \dfrac{\partial u}{\partial t} = \dfrac{1}{2}\dfrac{\partial^2 u}{\partial x^2} + |u|^\gamma \dot{W}, \ t \geq 0, \ x \in \mathbb{R} \\ u(0,x) = u_0(x), \end{cases} \tag{3.14}$$

where $u_0(x) \geq 0$. We expect the solution to be nonnegative, so that we can replace $|u|^\gamma$ by u^γ. The behavior of the solution to this equation, which has been the object of intense study, depends very much upon the value of the positive parameter γ. The case $\gamma = 1$ is easy and has already been considered in these notes. If $\gamma > 1$, then the mapping $r \to r^\gamma$ is locally Lipschitz, and there exists a unique strong solution, possibly up to an explosion time. C. Mueller has shown that the solution is strictly positive, in the sense that

$$u_0 \not\equiv 0 \Longrightarrow u(t,x) > 0, \ \forall t > 0, x \in \mathbb{R}, \ \mathbb{P} \text{ a.s.}$$

We shall consider here the case $\gamma < 1$.

3.7.1 The case $\gamma = 1/2$

In this case, the SPDE (3.14) is related to the super Brownian motion, which we now define. For a more complete introduction to superprocesses and for all the references to this subject, we refer the reader to [6]. Let $\mathcal{M}_d$ denote the space of finite measures on $\mathbb{R}^d$, and C_{c+}^d the space of C^2 functions from $\mathbb{R}^d$ into $\mathbb{R}_+$, with compact support. We shall use $\langle \cdot, \cdot \rangle$ to denote the pairing between measures and functions from C_{c+}^d.

Definition 3.13. The super Brownian motion is a Markov process $\{X_t, \ t \geq 0\}$ with values in $\mathcal{M}_d$ which is such that $t \to \langle X_t, \varphi \rangle$ is right continuous for all $\varphi \in C_{c+}^d$, and whose transition probability is characterized as follows through its Laplace transform

$$\mathbb{E}_\mu[\exp(-\langle X_t, \varphi \rangle)] = \exp(-\langle \mu, V_t(\varphi) \rangle), \quad \varphi \in C_{c+}^d,$$

where $\mu \in \mathcal{M}_d$ denotes the initial condition and $V_t(\varphi)$ is the function which is the value at time t of the solution of the nonlinear PDE

$$\begin{cases} \dfrac{\partial V}{\partial t} = \dfrac{1}{2}(\Delta V - V^2) \\ V(0) = \varphi. \end{cases}$$

Let us compute the infinitesimal generator of this diffusion.

If $F(\mu) = e^{-\langle \mu, \varphi \rangle}$,

$$\lim_{t \to 0} \frac{1}{t} \left(\mathbb{E}_\mu F(X_t) - F(\mu) \right) = \lim_{t \to 0} \frac{1}{t} \left(e^{-\langle \mu, V_t(\varphi) \rangle} - e^{-\langle \mu, \varphi \rangle} \right)$$

$$= -e^{-\langle \mu, \varphi \rangle} \lim_{t \to 0} \langle \mu, \frac{V_t(\varphi) - \varphi}{t} \rangle$$

$$= -\frac{1}{2} e^{-\langle \mu, \varphi \rangle} \langle \mu, \Delta \varphi - \varphi^2 \rangle$$

$$= \mathcal{G} F(\mu).$$

From this we deduce that if F has the form $F(X_t) = f(\langle X_t, \varphi \rangle)$, then

$$\mathcal{G} F(\mu) = \frac{1}{2} f'(\langle \mu, \varphi \rangle) \langle \mu, \Delta \varphi \rangle + \frac{1}{2} f''(\langle \mu, \varphi \rangle) \langle \mu, \varphi^2 \rangle.$$

Consequently, the process defined for $\varphi \in C_{c+}^d$ as

$$M_t^\varphi = \langle X_t, \varphi \rangle - \langle X_0, \varphi \rangle - \frac{1}{2} \int_0^t \langle X_s, \Delta \varphi \rangle ds$$

is a continuous martingale with associated increasing process

$$\langle M^\varphi \rangle_t = \int_0^t \langle X_s, \varphi^2 \rangle ds.$$

We have just formulated the martingale problem which the super Brownian motion solves. Let us show how this follows from our previous computations. We have that whenever $F(X_t) = f(\langle X_t, \varphi \rangle)$,

$$F(X_t) = F(X_0) + \int_0^t \mathcal{G} F(X_s) ds \quad \text{is a martingale.}$$

If we choose $f(x) = x$, we get that the following is a martingale

$$M_t^x = \langle X_t, \varphi \rangle - \langle X_0, \varphi \rangle - \frac{1}{2} \int_0^t \langle X_s, \Delta \varphi \rangle ds.$$

If we choose now $f(x) = x^2$, we get another martingale

$$M_t^{x^2} = \langle X_t, \varphi \rangle^2 - \langle X_0, \varphi \rangle^2 - \int_0^t \langle X_s, \varphi \rangle \langle X_s, \Delta \varphi \rangle ds$$

$$- \int_0^t \langle X_s, \varphi^2 \rangle ds.$$

Now applying Itô's formula to the first of the above two formulas yields

$$\langle X_t, \varphi \rangle^2 = \langle X_0, \varphi \rangle^2 + \int_0^t \langle X_s, \varphi \rangle \langle X_s, \Delta \varphi \rangle ds$$

$$+ \langle M^x \rangle_t + \text{martingale.}$$

Comparing the two last formulas gives

$$\langle M^x \rangle_t = \int_0^t \langle X_s, \varphi^2 \rangle ds.$$

Existence of a density and SBM-related SPDE

If $d \geq 2$, one can show that the measure X_t is a.s. singular with respect to Lebesgue measure. In contrast, if $d = 1$, the law of X_t is absolutely continuous w.r.t. Lebesgue measure. Define $u(t, \cdot)$ as the density of X_t. The formula for $\langle M^x \rangle_t$ implies that there exists a Gaussian random measure on $\mathbb{R}_+ \times \mathbb{R}$ such that

$$M_t^x = \int_0^t \int_{\mathbb{R}} \sqrt{u(s,x)} \varphi(x) W(ds, dx),$$

hence $u(t, x)$ is a (weak) positive solution of the SPDE

$$\begin{cases} \dfrac{\partial u}{\partial t} = \dfrac{1}{2} \dfrac{\partial^2 u}{\partial x^2} + \sqrt{u} \dot{W}, \ t \geq 0, \ x \in \mathbb{R} \\ u(0, x) = u_0(x). \end{cases} \tag{3.15}$$

Uniqueness in law

We now show that the law of the super Brownian motion is uniquely determined, which implies uniqueness in law for the SPDE (3.15).

From the Markov property, and the semigroup property of $\{V_t\}$, we deduce that

$$\mathbb{E}_\mu \left(e^{-\langle X_t, V_{T-t}(\varphi) \rangle} | \mathcal{F}_s \right) = \mathbb{E}_{X_s} \left(e^{-\langle X_{t-s}, V_{T-t}(\varphi) \rangle} \right)$$
$$= e^{-\langle X_s, V_{t-s}(V_{T-t}(\varphi)) \rangle}$$
$$= e^{-\langle X_s, V_{T-s}(\varphi) \rangle}.$$

We have just proved that $\{e^{-\langle X_t, V_{T-t}(\varphi) \rangle}, \ 0 \leq t \leq T\}$ is a martingale. Hence in particular

$$\mathbb{E}_\mu e^{-\langle X_T, \varphi \rangle} = e^{-\langle \mu, V_T(\varphi) \rangle},$$

which characterizes the transition probability of $\{X_t\}$, hence its law.

A construction of the SBM

We start with an approximation by a branching process.

- At time 0, let N particles have i.i.d. locations in $\mathbb{R}^d$, with the common law μ.
- At each time k/N, $k \in \mathbb{N}$, each particle dies with probability $1/2$ and gives birth to 2 descendants with probability $1/2$.

- On each interval $[k/N, (k+1)/N]$, the living particles follow mutually independent Brownian motions.

Denote by $N(t)$ the number of particles alive at time t, and Y_t^i the position of the i-th particle $(1 \le i \le N(t))$. Let $\{X_t^N\}$ denote the $\mathcal{M}_d$-valued process

$$X_t^N = \frac{1}{N} \sum_{i=1}^{N(t)} \delta_{Y_t^i}, \quad \langle X_t^N, \varphi \rangle = \frac{1}{N} \sum_{i=1}^{N(t)} \varphi(Y_t^i).$$

Theorem 3.14. $X^N \Rightarrow X$, as $N \to \infty$, where X is an SBM with initial law μ.

We shall not prove this theorem. We refer the reader to Etheridge [6].

Corollary 3.15. There exists a stopping time τ, with $\tau < \infty$ a.s., such that $X_\tau = 0$.

PROOF: The extinction time T of a branching process as described above satisfies, from a result due to Kolmogorov,

$$\mathbb{P}(T > t) = \mathbb{P}(NT > Nt) \simeq \frac{C}{Nt}.$$

Now with N independent such processes

$$\mathbb{P}(\sup_{i \le i \le N} T_i \le t) = \prod_{i=1}^{N} \mathbb{P}(T_i \le t) \simeq (1 - \frac{C}{Nt})^N \to e^{-C/t},$$

as $N \to \infty$. In other words, $\mathbb{P}(\tau > t) \simeq 1 - e^{-C/t}$. □

We will now show that whenever u_0 has compact support, the same is true with $u(t, \cdot)$, $\forall t > 0$. This follows from the next theorem.

Theorem 3.16. Let $\mu \in \mathcal{M}_d$ be such that $\mathrm{supp}\mu \subset B(0, R_0)$. Then, for all $R > R_0$,

$$\mathbb{P}(X_t(B(0, R)^c) = 0, \forall t \ge 0) = \exp\left(-\frac{\langle \mu, u(R^{-1} \cdot) \rangle}{R^2}\right),$$

where u is the positive solution of the PDE

$$\begin{cases} \Delta u = u^2, & |x| < 1; \\ u(x) \to \infty, & x \to \pm 1. \end{cases}$$

Corollary 3.17. Under the assumptions of the theorem,

$$\mathbb{P}_\mu \left(\cup_{t \ge 0} \mathrm{supp} X_t \text{ is bounded}\right) = 1.$$

PROOF: We have

$$\mathbb{P}_\mu \left(\cup_{t\geq 0} \mathrm{supp} X_t \text{ is bounded} \right)$$
$$= \mathbb{P}_\mu \left(\cup_{r>R_0} \{ X_t(B(0,r)^c) = 0, \ \forall t \geq 0 \} \right)$$
$$= \lim_{r\to\infty} \exp\left(-\frac{\langle \mu, u(r^{-1}\cdot)\rangle}{r^2} \right)$$
$$\geq \lim_{r\to\infty} \exp\left(-\frac{1}{r^2} [\sup_{|y|\leq R_0/r} u(y)] \mu(\mathbb{R}^d) \right)$$
$$= 1,$$

where we have used the theorem for the second equality. $\qquad\square$

Before we prove the theorem, we need one more lemma.

Lemma 3.18. $\forall t \geq 0$, $\varphi \in C_{c+}^d$, we have

$$\mathbb{E}_\mu \exp\left(-\int_0^t \langle X_s, \varphi\rangle ds \right) = \exp\left(-\langle \mu, u_t(\varphi)\rangle \right),$$

where $\{u_t(\varphi), \ t \geq 0\}$ is the positive solution of the nonlinear parabolic PDE

$$\begin{cases} \dfrac{\partial u}{\partial t} = \dfrac{1}{2}(\Delta u - u^2) + \varphi, & t \geq 0; \\ u(0) = 0. \end{cases}$$

Proof: Let $n \in \mathbb{N}$, $h = t/n$, $t_i = ih$.

$$\exp\left(-\int_0^t \langle X_s, \varphi\rangle ds \right) = \lim_n \exp\left(-\sum_{i=1}^n \langle X_{t_i}, h\varphi\rangle \right).$$

Now

$$\mathbb{E}_\mu \left(e^{-\langle X_{t_n}, h\varphi\rangle} | \mathcal{F}_{t_{n-1}} \right) = e^{-\langle X_{t_{n-1}}, V_h(h\varphi)\rangle},$$
$$\mathbb{E}_\mu \left(e^{-\langle X_{t_n}, h\varphi\rangle - \langle X_{t_{n-1}}, h\varphi\rangle} | \mathcal{F}_{t_{n-2}} \right) = \mathbb{E}_\mu \left(e^{-\langle X_{t_{n-1}}, V_h(h\varphi)+h\varphi\rangle} | \mathcal{F}_{t_{n-2}} \right)$$
$$= e^{-\langle X_{t_{n-2}}, V_h(V_h(h\varphi)+h\varphi)\rangle},$$

and iterating this argument, we find that

$$\mathbb{E}_\mu \exp\left(-\sum_{i=1}^n \langle X_{t_i}, h\varphi\rangle \right) = \exp\left(-\langle \mu, v_h(t)\rangle \right),$$

where v_h solves the parabolic PDE

$$\begin{cases} \dfrac{\partial v_h}{\partial t} = \dfrac{1}{2}(\Delta v_h - v_h^2), & ih < t < (i+1)h; \\ v_h(ih) = v_h(ih^-) + h\varphi \\ v_h(0) = 0. \end{cases}$$

In other words (here $P(t)$ stands for the semigroup generated by $\frac{1}{2}\Delta$)

$$v_h(t) = -\frac{1}{2}\int_0^t P(t-s)v_h^2(s)ds + h\sum_{0\le i:\, ih\le t} P(t-ih)\varphi$$

and letting n tend to $+\infty$ we have

$$u(t) = -\frac{1}{2}\int_0^t P(t-s)u^2(s)ds + \int_0^t P(t-s)\varphi ds.$$

PROOF OF THEOREM 3.16: Approximating the indicator function of the closed ball $B(0,R)$ by regular functions φ, and exploiting the fact that $t \to \langle X_t, \varphi\rangle$ is a.s. right continuous, as well as the monotone convergence theorem, we get that

$$\mathbb{P}_\mu(X_t(B(0,R)^c) = 0,\ \forall t \ge 0) = \mathbb{P}_\mu\left(\int_0^\infty X_t(B(0,R)^c)dt = 0\right)$$

$$= \lim_{\theta\to\infty}\mathbb{E}_\mu\left(\exp\left[-\theta\int_0^\infty X_t(B(0,R)^c)dt\right]\right)$$

$$= \lim_{\theta\to\infty}\lim_{n\to\infty}\lim_{m\to\infty}\lim_{T\to\infty}\mathbb{E}_\mu\left(\exp\left[-\int_0^T\langle X_t, \theta\varphi_{R,n,m}\rangle dt\right]\right)$$

$$= \lim_{\theta\to\infty}\lim_{n\to\infty}\lim_{m\to\infty}\lim_{T\to\infty}\exp\left[-\langle\mu, u_{n,m}(T,\cdot; R,\theta)\rangle\right],$$

where $\varphi_{R,n,m}$ is zero outside $[-m-1,-R]\cup[R,m+1]$, 1 on the interval $[-m,-R-1/n]\cup[R+1/n,m]$, increases and decreases linearly between 0 and 1; and $u_{n,m}(t,\cdot,R,\theta)$, by the preceding lemma, solves the parabolic PDE

$$\begin{cases}\dfrac{\partial v}{\partial t} = \dfrac{1}{2}(\Delta v - v^2) + \theta\varphi_{R,n,m},\ 0 \le t \le T,\\ v(0) = 0.\end{cases}$$

Now as $T \to \infty$, $u_{n,m}(T,\cdot,R,\theta) \to u_{n,m}(\cdot,R,\theta)$, which solves the PDE

$$-\Delta u_{n,m} + u_{n,m}^2 = 2\theta\varphi_{R,n,m},$$

and as $n, m \to \infty$, $u_{n,m}(\cdot,R,\theta) \to u(\cdot,R,\theta)$, a solution of

$$-\Delta u + u^2 = 2\theta 1_{|x|>R},$$

hence as $\theta \to \infty$, $u(\cdot,R,\theta) \to u(\cdot,R)$, a solution of

$$\begin{cases}-\Delta u + u^2 = 0,\quad |x| < R;\\ u(x) \to \infty,\quad x \to \pm R.\end{cases}$$

Since $u(x,R) = \frac{1}{R^2}u(\frac{x}{R})$, we finally get that

$$\mathbb{P}(X_t(B(0,R)^c) = 0,\ \forall t \ge 0) = \exp\left(-\frac{\langle\mu, u(R^{-1}\cdot)\rangle}{R^2}\right).$$

3.7.2 Other values of $\gamma < 1$

Mytnik [20] has proved that uniqueness in law holds if $1/2 < \gamma < 1$. Mueller and
Perkins [18] have proved that the compact support property is still true if $0 < \gamma < 1/2$.

3.8 A Reflected SPDE

In this section, we want first to study the following SPDE with additive white noise
and reflection

$$
\begin{cases}
\dfrac{\partial u}{\partial t} = \dfrac{\partial^2 u}{\partial x^2} + \eta + \dot{W}, \\[2mm]
u(0,x) = u_0(x), \quad u(t,0) = u(t,1) = 0, \\[2mm]
u \geq 0, \ \eta \geq 0, \ \displaystyle\int_0^\infty \int_0^1 u(t,x)\eta(\mathrm{d}t, \mathrm{d}x) = 0,
\end{cases}
\tag{3.16}
$$

where $u_0 \in C_0([0,1];\mathbb{R}_+)$. Without the measure η, the sign of the solution would
oscillate randomly. The measure η is there in order to prevent the solution u from
crossing 0, by "pushing" the solution upwards. The last condition says that the
pushing is minimal, in the sense that the support of η is included in the set where u
is zero. We formulate a precise definition.

Definition 3.19. A pair (u, η) is said to be a solution of equation (3.16) whenever
the following conditions are met:

1. $\{u(t,x), \ t \geq 0, \ 0 \leq x \leq 1\}$ is a nonnegative continuous and adapted process,
   such that $u(t,0) = u(t,1) = 0, \forall t \geq 0$.
2. $\eta(\mathrm{d}t, \mathrm{d}x)$ is an adapted random measure on $\mathbb{R}_+ \times [0,1]$.
3. For any $t > 0$, any $\varphi \in C_C^\infty([0,1])$, we have

$$
(u(t), \varphi) = (u_0, \varphi) + \int_0^t (u(s), \varphi'')\mathrm{d}s + \int_0^t \int_0^1 \varphi(x)W(\mathrm{d}s, \mathrm{d}x)
$$

$$
+ \int_0^t \int_0^1 \varphi(x)\eta(\mathrm{d}s, \mathrm{d}x).
$$

We have the following theorem (see [22]).

Theorem 3.20. If $u_0 \in C_0([0,1];\mathbb{R}_+)$, equation (3.16) has a unique solution.

PROOF: STEP 1 We first reformulate the problem. Let v denote the solution of the
heat equation with additive white noise, but without the reflection, i.e. v solves

$$
\begin{cases}
\dfrac{\partial v}{\partial t} = \dfrac{\partial^2 v}{\partial x^2} + \dot{W}, \\[2mm]
v(0,x) = u_0(x), \quad v(t,0) = v(t,1) = 0.
\end{cases}
$$

Defining $z = u - v$, we see that the pair (u, η) solves equation (3.16) if and only if z solves

$$
\begin{cases}
\dfrac{\partial z}{\partial t} = \dfrac{\partial^2 z}{\partial x^2} + \eta, \\[2mm]
z(0, x) = 0, \quad z(t, 0) = z(t, 1) = 0, \\[2mm]
z \geq -v, \ \eta \geq 0, \ \displaystyle\int_0^\infty \int_0^1 (z + v)(t, x)\eta(dt, dx) = 0.
\end{cases}
\tag{3.17}
$$

This is an obstacle problem, which can be solved path by path.

STEP 2 We construct a solution by means of the penalization method. For each $\varepsilon > 0$ let z_ε solve the penalized PDE

$$
\begin{cases}
\dfrac{\partial z_\varepsilon}{\partial t} = \dfrac{\partial^2 z_\varepsilon}{\partial x^2} + \dfrac{1}{\varepsilon}(z_\varepsilon + v)^-, \\[2mm]
z_\varepsilon(0, x) = 0, \quad z_\varepsilon(t, 0) = z_\varepsilon(t, 1) = 0.
\end{cases}
$$

It is easily seen that this equation has a unique solution in $L^2_{\text{loc}}(\mathbb{R}_+; H^2(0, 1)) \cap C(\mathbb{R}_+ \times [0, 1])$. Moreover, clearly z_ε increases, when ε decreases to 0. If z_ε and $\hat{z}_\varepsilon$ are solutions to the same equation, corresponding to v and $\hat{v}$ respectively, it is easy to show that

$$
\sup_{0 \leq t \leq T, 0 \leq x \leq 1} |z_\varepsilon(t, x) - \hat{z}_\varepsilon(t, x)| \leq \sup_{0 \leq t \leq T, 0 \leq x \leq 1} |v(t, x) - \hat{v}(t, x)|.
\tag{3.18}
$$

Let us show that

$$
w = z_\varepsilon - \hat{z}_\varepsilon - \|v - \hat{v}\|_{\infty, T} \leq 0,
$$

the same being true if we replace $z_\varepsilon - \hat{z}_\varepsilon$ by $\hat{z}_\varepsilon - z_\varepsilon$. The function w solves

$$
\begin{cases}
\dfrac{\partial w}{\partial t} = \dfrac{\partial^2 w}{\partial x^2} + \dfrac{1}{\varepsilon}(z_\varepsilon + v)^-, \\[2mm]
w(0, x) = -k, \quad w(t, 0) = w(t, 1) = -k,
\end{cases}
$$

where $k = \|v - \hat{v}\|_{\infty, T}$. If w reaches 0, then $z_\varepsilon \geq \hat{z}_\varepsilon + k$, hence $z_\varepsilon + v \geq \hat{z}_\varepsilon + \hat{v}$ and $(z_\varepsilon + v)^- \leq (\hat{z}_\varepsilon + \hat{v})^-$. In that case, the drift in the equation pushes w downwards, i.e. w remains negative between $t = 0$ and $t = T$. This intuitive argument can be justified by standard methods.

STEP 3 We let $z = \lim_{\varepsilon \to 0} z_\varepsilon$. We want to prove that z is continuous. If we replace v by a smooth obstacle v_n, then the difference between z_ε and $z_{n,\varepsilon}$ is dominated by $\|v - v_n\|_{\infty, T}$, and in the limit as $\varepsilon \to 0$,

$$
\|z - z_n\|_{\infty, T} \leq \|v - v_n\|_{\infty, T}.
$$

But it is known that when the obstacle v_n is smooth, z_n is continuous. Consequently z is the uniform limit of continuous functions, hence it is continuous.

STEP 4 Define

$$
\eta_\varepsilon(dt, dx) = \varepsilon^{-1}(z_\varepsilon + v)^-(t, x)dtdx.
$$

Since $\eta_\varepsilon = \frac{\partial z_\varepsilon}{\partial t} - \frac{\partial^2 z_\varepsilon}{\partial x^2}$, by integration by parts we deduce that for any smooth function ψ of (t,x) which is zero whenever $x = 0$ or $x = 1$,

$$\langle \eta_\varepsilon, \psi \rangle = -\int_0^\infty (z_\varepsilon, \frac{\partial \psi}{\partial t} + \frac{\partial^2 \psi}{\partial x^2}) dt,$$

hence $\eta_\varepsilon \to \eta$ in the sense of distributions, as $\varepsilon \to 0$. The limit distribution is nonnegative, hence it is a measure, which satisfies

$$\langle \eta, \psi \rangle = -\int_0^\infty (z, \frac{\partial \psi}{\partial t} + \frac{\partial^2 \psi}{\partial x^2}) dt.$$

Now the support of η_ε is included in the set $\{z_\varepsilon + v \le 0\}$, which decreases as $\varepsilon \to 0$. Hence the support of η is included in $\{z_\varepsilon + v \le 0\}$ for all $\varepsilon > 0$. Consequently for all $T > 0$,

$$\int_0^T \int_0^1 (z_\varepsilon + v) d\eta \le 0.$$

The same is true with z_ε replaced by z by monotone convergence. Hence

$$\int_0^T \int_0^1 (z + v) d\eta = 0.$$

STEP 5 If the solution were in $L^2_{\text{loc}}(\mathbb{R}_+; H^1(0,1))$, then the uniqueness proof would follow a very standard argument; since if (z, η) and $(\bar{z}, \bar{\eta})$ are two solutions,

$$\int_0^T \int_0^1 (z - \bar{z}) d(\eta - \bar{\eta}) \le 0.$$

Since the above regularity does not hold, one needs to implement a delicate regularization procedure, which we will not present here.   □

The reflected white noise driven SPDE is related to the following SPDEs with singular drift

$$\begin{cases} \dfrac{\partial u}{\partial t} = \dfrac{\partial^2 u}{\partial x^2} + \dfrac{c}{u^\alpha} + \dot{W}, \\ u(0,x) = u_0(x), \quad u(t,0) = u(t,1) = 0. \end{cases}$$

It has been shown that the solution of such an equation remains strictly positive if $\alpha > 3$, and has positive probability of hitting 0 if $\alpha < 3$. The case $\alpha = 3$ is the most interesting, since the solution might touch zero at isolated points, and one can define the solution for all time. Now, consider the SPDE with singular drift

$$\begin{cases} \dfrac{\partial u}{\partial t} = \dfrac{\partial^2 u}{\partial x^2} + \dfrac{(\delta - 1)(\delta - 3)}{8u^3} + \dot{W}, \\ u(0,x) = u_0(x), \quad u(t,0) = u(t,1) = 0, \end{cases}$$

where $\delta > 3$. It can be shown that the solution of this SPDE converges to the above reflected SPDE, as $\delta \to 3$. L. Zambotti has shown in [31] that the solution to these equations are ergodic, and explicitly computed their invariant measure (including in the case $\delta = 3$), with respect to which the process is reversible. It is the law of the δ-Bessel bridge, i.e. that of the δ-Bessel process, conditioned to be at 0 at time 1. The δ-Bessel process is the solution of the one-dimensional SDE

$$dX_\delta(t) = \frac{\delta - 1}{2X_\delta(t)} dt + dW(t), \; X_\delta(0) = 0.$$

In the case where δ is an integer, it has the same law as the norm of the δ-dimensional Brownian motion.

Moreover, Dalang, Mueller and Zambotti [3] have given precise indications concerning the set of points where the solution hits zero. This set is decreasing in δ. For $\delta = 3$, with positive probability there exist three points of the form $(t, x_1), (t, x_2), (t, x_3)$ where u is zero, and the probability that there exist 5 points of the same form where u hits zero is zero. For $4 < \delta \leq 5$, there exists one such point with positive probability, and two such points with zero probability. For $\delta > 6$, the probability that there exists one point where u hits zero is zero.

Finally, let us mention that white noise-driven reflected SPDEs with a solution-dependent diffusion coefficient multiplying the noise have been studied by Donati-Martin and Pardoux [5].

References

1. V. Bally and E. Pardoux, Malliavin calculus for white noise driven parabolic SPDEs, *Potential Analysis* **9**, 27–64, 1998.
2. L. Bertini and G. Giacomin, Stochastic Burgers and ZPZ equations for particle systems, *Comm. Math. Phys.* **183**, 571–607, 1997.
3. R. Dalang, C. Mueller and L. Zambotti, Hitting properties of parabolic spde's with reflection, *Ann. Probab.* **34**, 1423–1450, 2006.
4. G. Da Prato and J. Zabczyk, *Stochastic equations in infinite dimension*, Cambridge Univ. Press 1995.
5. C. Donati-Martin and E. Pardoux, White noise driven SPDEs with reflection, *Prob. Theory and Rel. Fields* **95**, 1–24, 1993.
6. A. Etheridge, *Introduction to superprocesses*, Memoirs of the AMS, 2000.
7. T. Funaki and S. Olla, Fluctuation for the $\nabla \phi$ interface model on a wall, *Stoch. Proc. and Applic.* **94**, 1–27, 2001.
8. M. Gubinelli, P. Imkeller and N. Perkowski, Paracontrolled distributions and singular PDEs, *Forum of Mathematics, Pi* **3**, 2015.
9. I. Gyöngy and E. Pardoux, Weak and strong solutions of white-noise driven parabolic SPDEs, *Unpublished manuscript*, 1992.
10. M. Hairer, A theory of regularity structures, *Inventiones mathematicae* **198**, 269–504, 2014.
11. M. Hairer and P. Friz, *A Course on Rough Paths, with an introduction to regularity structures*, Springer, 2014.
12. N.V. Krylov, A relatively short proof of Itô's formula for SPDEs and its applications, *Stochastic Partial Differential Equations: Analysis and Computations* **1**, 152–174, 2013.
13. N.V. Krylov and B.L. Rozovsky, Stochastic evolution systems, *Russian Math. Surveys* **37**, 81–05, 1982.
14. J.L. Lions, *Quelques méthodes de résolution des problèmes aux limites non linéaires et applications*, Dunod, 1969.
15. P.L. Lions and P.E. Souganidis, Fully nonlinear stochastic partial differential equations: nonsmooth equations and applications, *Comptes Rendus de l'Acad. Sciences, Série 1, Mathematics* **327**, 735–741, 1998.
16. J. Mattingly and E. Pardoux, Malliavin calculus for the stochastic 2D Navier–Stokes equation, *Comm. Pure and Appl. Math* **59**, 1742–1790, 2006.
17. M. Métivier, *Semimartingales*, de Gruyter 1982.
18. C. Mueller and E. Perkins, The compact support property for solutions to the heat equation with noise, *Probab. Theory and Rel. Fields* **93**, 325–358, 1992.
19. C. Mueller and E. Pardoux, The critical exponent for a stochastic PDE to hit zero, *Stochastic analysis, control, optimization and applications*, 325–338, Birkhäuser 1999.
20. L. Mytnik, Weak uniqueness for the heat equation with noise, *Ann. Probab.* **26**, 968–984, 1998.
21. D. Nualart, *The Malliavin calculus and related topics*, Springer 1995.
22. D. Nualart and E. Pardoux, White noise driven quasilinear SPDEs with reflection, *Probab. Theory and Rel. Fields* **93**, 77–89, 1992.

© The Author(s), under exclusive license to Springer Nature Switzerland AG 2021
É. Pardoux, *Stochastic Partial Differential Equations*,
SpringerBriefs in Mathematics, https://doi.org/10.1007/978-3-030-89003-2

23. E. Pardoux, Equations aux dérivées partielles stochastiques monotones, Thèse, Univ. Paris–Sud, 1975.
24. E. Pardoux, Stochastic partial differential equations and filtering of diffusion processes, *Stochastics* **3**, 127–167, 1979.
25. E. Pardoux, Filtrage nonlinéaire et équations aux dérivées partielles stochastiques associées, in *Ecole d'été de Probabilités de Saint Flour XIX*, Lecture Notes in Math. **1464**, 67–163, Springer 1991.
26. E. Pardoux and A. Răşcanu, *Stochastic Differential Equations, Backward SDEs, Partial Differential Equations*, Stochastic Modelling and Applied Probability **69**, Springer 2014.
27. E. Pardoux and T. Zhang, Absolute continuity of the law of the solution of a parabolic SPDE, *J. of Funct. Anal.* **112**, 447–458, 1993.
28. B. Rozovsky, *Evolution stochastic systems*, D. Reidel 1990.
29. D.W. Stroock and S.R.S. Varadhan, *Multidimensional Diffusion Processes*, Classics in Mathematics, Springer 2006.
30. J. Walsh, An introduction to stochastic partial differential equations, *Ecole d'été de Probabilités de Saint Flour XIV*, Lecture Notes in Math. **1180**, 265–439, Springer 1986.
31. L. Zambotti, Integration by parts on δ-Bessel bridges, $\delta > 3$, and related SPDEs, *Ann. Probab.* **31**, 323–348, 2003.
32. B.Z. Zangeneh, Semilinear stochastic evolution equations with monotone nonlinearities, *Stochastics* **53**, 129–174, 1995.

Index

Printed in the United States
by Baker & Taylor Publisher Services